T0296562

CO-ORDINATION OF GALACTIC RESEARCH

INTERNATIONAL ASTRONOMICAL UNION

SYMPOSIUM No. 1

HELD AT GRONINGEN, 22–27 JUNE 1953

CO-ORDINATION OF GALACTIC RESEARCH

EDITED BY

A. BLAAUW

Yerkes Observatory

CAMBRIDGE

AT THE UNIVERSITY PRESS

1955

CAMBRIDGE
UNIVERSITY PRESS

University Printing House, Cambridge CB2 8BS, United Kingdom

Cambridge University Press is part of the University of Cambridge.

It furthers the University's mission by disseminating knowledge in the pursuit of education, learning and research at the highest international levels of excellence.

www.cambridge.org
Information on this title: www.cambridge.org/9781316612705

© Cambridge University Press 1955

This publication is in copyright. Subject to statutory exception and to the provisions of relevant collective licensing agreements, no reproduction of any part may take place without the written permission of Cambridge University Press.

First published 1955
First paperback edition 2016

A catalogue record for this publication is available from the British Library

ISBN 978-1-316-61270-5 Paperback

Cambridge University Press has no responsibility for the persistence or accuracy of URLs for external or third-party internet websites referred to in this publication, and does not guarantee that any content on such websites is, or will remain, accurate or appropriate.

PREFACE

A conference on co-ordination of galactic research was held near Groningen, Netherlands, in June 1953. The report given in the following chapters summarizes the most essential parts of the discussions. In accordance with a decision taken during the conference, no attempt has been made to reproduce the various speeches given by the introductory speakers. It was considered more desirable to have a report with the relevent remarks arranged roughly according to the scheme of the memorandum that had been drawn up before the conference as a basis for the discussions. Many topics have been discussed on various occasions and we have attempted to refer to them in the general context of the subject. No attempt has been made to indicate consistently which of the participants was responsible for the opinions expressed in the report—this would have been impossible anyhow, as usually such opinions arose as the result of general discussions of the subject.

Some of the introductory papers have in the meantime been published or are fairly well covered by certain publications. Parenago's introductory speech on recent work of the Soviet astronomers has appeared in *Astronomical Newsletters*, Nos. 71 and 73. In other cases reference to such publications is given in the lists at the end of the report.

In the discussions, the terms 'halo', 'nucleus' and 'disk' are used to indicate different parts of the Galaxy. These general regions are not defined more precisely. Their introduction proved very useful, and one might rather say that their more exact description is one of the problems of galactic research.

The present report was drawn up by the undersigned, as secretary of the Organizing Committee, and submitted for changes and additions to all who had participated at the conference.

A. BLAAUW

LEIDEN OBSERVATORY, 1953

YERKES OBSERVATORY, 1954

PREFACE

CONTENTS

INTRODUCTION

The initiative in the organization of the conference was taken by the President of Commission 33 of the International Astronomical Union and financial aid was received from U.N.E.S.C.O. At the invitation of Dr P. J. van Rhijn, Director of the Kapteyn Astronomical Laboratory, Groningen, the meeting was held in the estate 'Vosbergen' near the city of Groningen and owned by the University of Groningen. The organizing committee consisted of J. H. Oort (Chairman), W. Baade, B. J. Bok, Ch. Fehrenbach, B. Lindblad, W. W. Morgan, P. P. Parenago, and A. Blaauw (Secretary), all of whom attended the conference. The other participants, who were invited either because they represented institutions which might take part in future galactic research, or because of the character of their research, were V. A. Ambartsumian, W. Becker, P. Couderc (representing the Commission for the Carte du Ciel), G. Haro, O. Heckmann, H. Spencer Jones, B. V. Kukarkin, J. J. Nassau, P. Th. Oosterhoff, L. Plaut (local Secretary), J. M. Ramberg, C. Schalen, J. Schilt, R. H. Stoy, B. Strömgren, P. J. van Rhijn. V. Kourganoff, P. G. Kulikovsky and O. A. Melnikov were present as interpreters.

Purpose and character of the conference

The following is extracted from a circular letter, sent to the participants before the beginning of the conference:

During the first third of this century an important concentration of work on galactic structure and motions has been promoted by the Plan of Selected Areas, initiated in 1906 by Kapteyn. Although the scheme outlined in 1906 has not lost its significance, it is widely felt that further research into structure and dynamics of the Galaxy should be extended beyond the original Plan. At the same time it is felt by many that some kind of co-ordination of effort remains highly desirable, if only because the value of many observations is greatly enhanced if the data can be combined with other data for the same stars.

Several observatories have asked for suggestions with regard to future work on galactic structure. The recent advent of several large Schmidt telescopes furnished with objective prisms, red-sensitive plates, etc., brings forward with some urgency for each of the observatories concerned the problem of how to organize and restrict the work with telescopes that can produce much more than can be measured and discussed by the existing observatories. The question arises, whether it is desirable to formulate a new plan of attack, and whether such a plan should also include recommendations for concentrated work on particular regions. If the answer to the latter is affirmative, which regions should be selected and what data would be most urgent?

Before such questions can be fruitfully considered it is desirable to try to formulate the principal problems in which we are most interested. The emphasis will no longer be confined to general overall structure, but more on structural features displayed by special types of stars, on relations between kinematical and distributional features, on spiral structure and other phenomena connected with the evolution of stars and the galactic system. It is clear that in all these problems motions will be equally important as space distribution.

It was the opinion of the Organizing Committee, that the conference would stand the best chance of becoming fruitful if, instead of progress reports, only a very limited number of carefully prepared introductions would be given, dealing mainly with the practical possibilities for galactic research offered by different types of observation. A list of these introductions is given below.

As was to be expected, the conference proved to be very useful for the formulation of the principal present problems. Concrete proposals for research with detailed specifications have been made only for a few of the many branches of galactic research, mainly for faint proper motion work and meridian observations. In some fields, it turned out that more experimental work will have to be done before we can embark upon a large-scale observing programme. In so far as the discussion stayed on the level of a general orientation of the main problems and the practical possibilities of attacking them, the conference served as a basis for more specific planning in the future.

Formation of new sub-committees in the I.A.U.

The need for continued consultation on galactic research was expressed by various participants during and after the meeting. Another meeting, perhaps about the time of the I.A.U. meeting in 1955, seems already desirable. Meanwhile, in order to guarantee continued co-ordination of activities in galactic research, a continuation committee for co-ordination of observational programmes was installed, after the conference, by the Executive Committee of the I.A.U., in accordance with a resolution adopted during the conference. This will be a sub-commission to Commission 33 and have the following members: Baade, Blaauw, Lindblad, Oort, Parenago, and van Rhijn. It is hoped, that this sub-commission will become a centre for information on current programmes, and an intermediary for exchange of ideas and suggestions wherever these are wanted by astronomers. The chairman of this sub-committee will be the chairman of Commission 33 (at present Dr J. H. Oort, Leiden Observatory) and the secretary will be A. Blaauw (Yerkes Observatory).

Another committee, sub-commission to Commission 27 (Variable Stars) has been formed by the Executive Committee, in order to organize the further work on variable-star surveys as described in section A (2). Its present members are Baade, Kukarkin, Oosterhoff, and the secretary of the above mentioned sub-commission of Commission 33.

Introductory papers

Before the discussions started on the various subjects reported below, the following introductions, mostly dealing with observational possibilities, were given:

W. Baade: Summary of results of recent extragalactic research.

P. P. Parenago: Outline of present galactic research in the Soviet Union.

B. Lindblad and C. Schalén: Work on stellar spectra and photometry at Stockholm and Uppsala.

W. W. Morgan: Spectroscopic observational possibilities for galactic research.

W. Becker: Recent results on three-colour photometry.

B. Strömgren: Recent developments of multi-colour photometry.

Ch. Fehrenbach: Objective prism radial velocity determinations.

O. A. Melnikov, C. Schalén, B. Lindblad and B. J. Bok: Communications on experimental work on the same subject.

J. J. Nassau: Possibilities and recent results based on observations in the infra-red.

J. H. Oort, W. W. Morgan, B. J. Bok: Recent results on spiral structure.

References in the following pages are to be found at the end of this publication on pp. 49–51.

A (1). OVERALL STRUCTURE: NUCLEAR REGION AND HALO

The objects from which the structure and the kinematical properties of the nuclear region and the halo can be derived are in general those classified as population II objects: the RR Lyrae variables, population II Cepheids, RV Tauri variables, long-period variables with periods near 200 days, novae, planetary nebulae, bright red giants as found in globular clusters, and the blue stars found near the galactic poles which are probably similar to those found in globular clusters.

Although the separation of halo and nuclear region is somewhat arbitrary, it is convenient for the discussions to use those terms. Some differences in populations of the two regions are indicated by the observations, and the observational approach presents different problems.

WORK ON VARIABLE STARS IN THE HALO

Experience gained through large surveys of variable stars (Harvard, Leiden, Sonneberg) and the statistical investigations based on their results have shown that this is a very promising field of research.[1] The aim of an extension of this work is more precise knowledge of the density distribution in the halo and near the galactic centre.

Large-scale structure of the halo

From all the evidence available at present it seems that the distribution of the population II objects is symmetrical with respect to the axis of rotation of the Galaxy. (Distribution of globular clusters in the Galaxy, distribution of population II objects in extra-galactic nebulae.) The first task of future work, therefore, will be to determine the way in which the density varies with the distance from the galactic centre, as measured along the galactic plane, and with the distance from this plane. The determination of possible deviations from rotational symmetry is a problem for the more distant future. This assumption implies that it is not necessary to observe in all galactic longitudes. The following proposal was made by Baade.

For the determination of the surfaces of equal density in the halo, we can restrict ourselves to determining the density distribution in a cross-section, perpendicular to the galactic plane and going through the Sun and the axis of rotation. The fields to be studied then will be at galactic longitudes 147° and 327°. The number and size of the fields depends on the number of objects to be expected and on the number of times one wants to cross a certain surface of equal density for the determination of its location. The surveying (observing, blinking) and the determination of light curves and apparent magnitudes involves a large amount of work. It is proposed that, to begin with, three fields be chosen, centred at approximately

$l=327°$, $b=$ about 20° (latitude as close to the nucleus as interstellar absorption permits);
$l=327°$, $b=45°$;
$l=147°$, $b=$ somewhere about 10°.

The ideal instrument for such a survey would be the 48-inch Palomar Schmidt, which may become free in the next few years after the sky survey is concluded. It gives fields of $7° \times 7°$ free from vignetting and can easily reach the 20th photographic magnitude in 10-min. exposures. This provides an ample margin to make sure that the survey will be complete up to 17·5 median magnitude corrected for absorption, i.e. distances up to 30 kiloparsecs for the RR Lyrae variables—even if the absorption in one of the fields should amount to one or one and a half magnitudes. In the case of the two latter fields it is very probable that regions of heavy or irregular absorption can be avoided. As to the first, it is a matter of searching the sky survey plates in this general direction before deciding on the most suitable co-ordinates.

The question has been raised if this new survey is justified in view of the large amount of work which has been done already in previous surveys.[1] The opinion of most of the participants at the conference was that little duplication will be involved because the new survey extends to fainter limiting magnitudes than the previous ones which, moreover, are mostly much nearer to the galactic plane. The most important feature, however, will be that strong emphasis would be placed upon completeness of the survey up to apparent magnitude 17·5. Exhaustive studies of variable-star fields at the intermediate galactic latitudes, up to this magnitude, have not been made and thus the extension of work in existing surveys would be considerable anyway. Another important consideration is that the existing surveys are made with telescopes which require fairly large field corrections, so that the limiting magnitude varies considerably over the field.

The exhaustive search for variables is a tedious task and has to be done by an experienced investigator who fully realizes the importance of searching a field very carefully, even by the time it is almost exhausted and only a few new discoveries are to be expected. The completeness factors to be applied to the final number of variables found should be so well determined that their uncertainty does not affect the final discussion. In every field, 300 plates should be taken for the determination of periods and light curves, and thirty pairs of these plates should be blinked. All RR Lyrae variables, long-period variables and semi-regular variables should be studied. From experience in this sort of work, Baade thinks that the variables of small amplitudes will be difficult to detect if they are more than five magnitudes brighter than the plate limit. For the brighter stars we need supplementary surveys (see pp. 7 and 8). Blue plates are recommended. There is strong evidence that the composition of the halo is similar to that of the globular clusters and in these we find the red variables at maximum to be 1·5 magnitudes brighter than the RR Lyrae variables. Therefore, if the survey will be complete for the RR Lyrae variables, it will also be so for the red variables. The use of photo-visual or red plates would reduce the amplitudes of the variables and considerably diminish the chances of discovery. This latter experience was also reported by Bok.

The amount of work required by this new survey can be roughly estimated on the basis of the following figures provided by Dr Plaut:

The time needed to blink ninety pairs of plates is hard to predict, but may be of the order of 2000 hours.

The total number of RR Lyrae variables to be expected in the 48-inch Schmidt field nearest to the galactic centre is about 200. The time needed to estimate the brightness of those 200 RR Lyrae variables on 300 plates and to determine periods and light curves is about 6000 hours.

The number of long-period and irregular variables is difficult to estimate. From the best evidence available one would guess that they are less numerous than the RR Lyrae variables. In any event the complete study of these objects will be less time-consuming than that of the RR Lyrae variables.

The total amount of time required for the field nearest to the galactic centre will thus be about 10,000 hours, and for the three fields together we would expect something of the order of 15,000 hours or six man-years.

Experiments to develop a more rapid process of finding the variables than by regular blinking or comparing negative and positive plates have not yet been successful.

Since the conference was held, it has become likely that Dr Plaut will be able to undertake the blinking, and that at least part of the further work will also be carried out at the Kapteyn Laboratory.

The photometric scale

For the above programme as well as for those mentioned below the precise reduction of the observations to a well defined photometric system is of fundamental importance. This phase of variable star work has been neglected too much in the past. It is strongly recommended that the sequences transferred to the variable star fields represent with very little tolerance the international system. If this condition is not satisfied, the results of the variable star statistics cannot be used for the accurate measurement of the dimensions and structure of the Galaxy.

An excellent basis for the transfer of the photometric scales will be the sequences in nine Selected Areas established by Baum. This work, which includes photo-electric observations of photographic magnitudes and colour indices from 9 to 19·5 magnitude will be completed about the time this report is published. In zero point and scale they will be copies of the polar sequence, but the internal accuracy will be much higher. Their transfer to the variable star fields, which may be either photographic or photo-electric, will be technically much easier than that of the polar sequence. The list of these precision areas follows:

S.A.	R.A.	Dec.	l	b
68	$0^h\ 11^m$	+15°	79°	−47°
94*	2 51	0	143	−48
71	3 11	+15	135	−34
51	7 24	+30	157	22
54	10 24	+30	168	60
57	13 4	+30	25	85
107*	15 34	0	334	40
61	16 59	+30	19	34
89	21 8	+15	33	−22

* For transfers to southern hemisphere.

Soviet survey of variable stars in Kapteyn's Areas

From communications by Dr B. V. Kukarkin we derive the following. The Moscow and Engelhardt (Kasan) Observatories have started (in 1948) a survey of variable stars in the Kapteyn Selected Areas from the equator up to the North Pole. The limiting magnitude is about 17·5 and the areas studied cover about 100 square degrees each. For every area, thirty pairs of plates will be blinked, and completeness will be achieved up to about $m = 17$. The instruments used are a 38 cm. Schmidt and a 40 cm. astrograph.

This programme will be a very valuable addition to the one outlined above. It does not penetrate to as large distances, but will permit a much more detailed study of the nearer regions, especially with regard to the different behaviour of various sub-groups among the RR Lyrae and long-period variables, and the properties of the uncommon types of variables, as for instance the RV Tauri stars.

General survey of the brightest variables

It is generally considered desirable, that in addition to the surveys mentioned above, a complete search over the whole sky be made for the stars brighter than, say, magnitude 12·0. This will provide information on the local density of RR Lyrae and long-period variables which are typical halo-objects. We also want to study the kinematical properties of these objects. This survey, therefore, should be supplemented by measurements of proper motions and radial velocities (see also Sections C (1) and C (4)).

Such a survey should be made with an instrument of short focal length, so that the density of the stellar background on the plates is sufficient for the blinking. Perhaps a meteor camera, if free of vignetting, would be the suitable instrument. Again, as in the other surveys, completeness is required; some of the stars discovered will be known already as variables and have known elements of the light variation (Harvard, Sonneberg, Odessa, etc.).

In this connexion it may be mentioned that another project under way in the Soviet Union (Odessa, and other observatories) is the sky patrol between declination minus 45° and the North Pole. Its purpose is to determine the complete characteristics (periods, light curves) of the *known* variables, brighter than 12·0 photographic magnitude. As Dr Bok pointed out, the Harvard sky patrol plates will be available for further study of the brighter variables to anybody interested in using them. Dr Kukarkin suggested that the trained staff of the three Soviet institutions working on variable stars may help in studying variable star plates obtained elsewhere.

LOW LUMINOSITY BLUE STARS IN THE HALO

To the surveys of variable stars one might add a survey of the blue stars of population II of the kind which have been discovered near the North Pole by Humason and Zwicky. Although the character of the stars found in their survey has not yet become quite clear, it seems highly probable that they have absolute magnitudes around $M=0$ or fainter.[1] The brighter ones—$m_{pg} < 12$—could be picked up from objective prism plates as in the case of the star observed at Cleveland.[2] For a general survey extending to fainter magnitudes, red and blue exposures with one of the existing

Schmidt telescopes could provide the necessary material. An experimental programme for finding the stars, based on simultaneous ultra-violet and yellow exposures with a small Schmidt camera, is now under way at the Yerkes and McDonald Observatories. The blue stars are expected to stand out as conspicuous objects on these plates. This method may prove to be more efficient than that of the objective prism spectra. The surveys should be followed up with photo-electric observations of the colours and slit spectra of the stars found.

NUCLEAR REGION

Radio surveys

The most accurate determination of the position of the galactic nucleus follows from radio surveys at decimetre and metre wave-lengths and from observations of the 21-cm. line of interstellar hydrogen. A very interesting question is, how steeply the intensity falls from the nucleus towards its immediate surroundings. The solution of this problem requires instruments of high resolving power.

Variable stars

Baade[1] has investigated the frequency of the variable stars in a small region of 16′ radius, $l=328°.2$, $b=-4°.3$ up to $m=20{\cdot}0$, and from the distribution in apparent magnitude he was able to determine the distance of the galactic centre and the density of the RR Lyrae variables for distances from 0·6 to 2·6 kiloparsecs from the nucleus. A similar investigation based on plates of a region at latitude $-7°.3$ near the galactic centre, observed at the Radcliffe Observatory, will probably be made at the Leiden Observatory. An investigation at galactic latitude $-20°$, longitude 307° down to the 17th magnitude had been published earlier by Shapley.[2] The region observed by Baade was picked out after careful scanning of the nuclear regions on films taken with the Palomar 18-inch Schmidt telescope. A field equally close to the galactic centre and probably somewhat more homogeneous is the region just south of Co $-29°14299$ ($l=329°.2$, $b=-4°.1$). The important criterion for the selection is not only sufficiently low absorption, but the absorption should also be uniform over the field and wisps of dark clouds which are frequent in these directions must be avoided. This condition is much easier to meet in somewhat higher latitudes and a second Pretoria field, centred on Co $-28°14334$ ($l=331°.7$, $b=-6°.5$) should be nearly ideal. Use of red or infra-red plates which would penetrate the absorption better is not to be preferred because in red light enormous numbers of faint stars come up which not only complicate the search for variables but are the cause of too much overlapping. There are two important additional reasons why red and infra-red plates should be avoided in investigations of the variable stars of the nuclear region: the small amplitudes of the RR Lyrae variables and Cepheids in the red and the complete lack of faint red and infra-red standard magnitudes.

Because of the high star densities in the nuclear region of our Galaxy and the strong overlying absorption large reflectors of sufficiently great focal length are required for the investigations just described. Large Schmidt telescopes with their relatively short focal lengths should become useful in latitudes greater than 7°.

Baade's field was centred on the globular cluster NGC 6522, which is imbedded in the centre region of our Galaxy at a distance of 10·1 kiloparsecs from the Sun, and he inferred the absorption for his field from the observed colour excess of NGC 6522. Obviously a determination of the absorption from the reddening of a number of the RR Lyrae variables in the field is highly desirable and attempts have been made both on Mount Wilson and Palomar Mountain to determine photo-electrically the colour excesses of a few of these variables. These attempts failed because the Sagittarius region is too far south for such precise measures. It is hoped that they can be provided in the future by observations from the southern hemisphere. Baade thinks there is no way of allowing for variable absorption over a field in a statistical investigation. It might be that very close to the centre even a smaller field than the one used by Baade would be sufficient because of the higher frequency of RR Lyrae variables to be expected there.

At distances of about 10° from the galactic centre, particularly in negative latitudes where the absorption is lower than at positive latitudes, a region may be found with sufficiently low and uniform absorption and large enough to give statistically significant numbers of the variables. This is a matter of further searching. If fields sufficiently large can be found in this latitude, large Schmidt telescopes undoubtedly will be the proper instruments. Investigation of these directions should give information on the densities at distances $\geqslant 1500$ parsecs from the nucleus itself. According to Bok, Schmidt plates taken with the ADH telescope show that at negative latitudes about $-7°$ the absorption becomes uniform.

Other types of variable stars in the nuclear region were discussed only briefly. In the Cleveland infra-red survey variables are picked out on the basis of the presence of the vanadium oxide bands. The possibilities of this work for a general survey of long-period variables are being investigated. Near the nucleus in the region investigated by Baade, the number of Mira variables with periods about 200 days and that of the variables with periods between 60 and 150 days and amplitudes about one magnitude, is about half of the RR Lyrae variables, and an equal number of semi-regular and irregulars were found. However, it is quite possible that they are relatively more numerous farther from the centre. It is to be borne in mind that the RR Lyrae variables discovered by Baade near the nucleus have shorter periods in general than those more distant from the centre, and seem to form a somewhat different class.

Dr Kulikovsky reported on work done by Vorontsov-Velyaminov and his associates on the distribution of long-period variables, and drew attention to an apparent concentration of these objects in the direction $l = 335^\circ$, $b = -7^\circ$. Whether indeed such concentrations are real or not, or are caused by local transparency of the interstellar medium, remains to be investigated.

Planetary nebulae

Minkowski's[1] and Henize's surveys of planetary nebulae are being supplemented by a new survey with the Schmidt telescope of the Tonantzintla Observatory. This survey covers the central region between approximately galactic longitudes 305° and 345° and latitudes ± 15°. Until now 437 H_α emission objects were found. Of these 121 correspond to previously discovered planetaries while sixty-seven are classified as new planetaries and forty-eight as probable ones, small diffuse nebulae or novae. The majority of the rest are probably peculiar emission stars. A general programme of checking the emission objects is planned at Tonantzintla.

The determination of radial velocities of the planetary nebulae would add very important information on the dynamics in the nuclear region. Two projects are already under way: Dr Mayall of the Lick Observatory is observing the spectra of all planetaries north of −25° which are not too heavily obscured, while Dr Minkowski at Mount Wilson and Palomar will observe the faint planetaries in the nuclear region. There is some overlap of the two programmes in order to check for systematic errors in the radial velocities of the two series. It is expected that the Radcliffe Observatory will observe planetary nebulae in the remaining part of the sky.

Infra-red survey of M giants

The work undertaken at the Warner and Swasey Observatory,[2] and recently extended to the southern hemisphere in collaboration with the Tonantzintla Observatory, will give important information on the space distribution of the late M giants. On plates of the Sagittarius cloud, Nassau classified between 300 and 400 late M's per square degree in the direction $l=330°$, $b=-9°$ and a somewhat smaller number at $l=326°$, $b=-8°$. The limiting infra-red magnitude of these counts is about 12·7. As the infra-red absorption in this direction is of the order of one to two magnitudes and the mean infra-red luminosity between -3 and -4, the survey reaches at least the distance of the nucleus. Thus, from these objects another measure of the density distribution in the central part of the Galaxy may result. The results announced by Nassau are part of a general survey at galactic latitude $-8°$ in a belt of $22°$ in longitude symmetrical with respect to the direction of the centre, for which plates are being taken by Haro.

Novae survey

The strong concentration of the novae to the galactic nucleus makes them interesting objects for the study of the nuclear region. A basic difficulty in the interpretation of the results from any survey will be the uncertainty in the interstellar absorption. There seems no means of evaluating this. However, important problems may be solved without exact knowledge of the locations in space. Examples are: the frequency of the novae phenomenon in our Galaxy, the apparent (projected) concentration to the galactic nucleus, the shape of the light curves and their correlation with spectral peculiarities. Apart from the interest of these investigations for the study of the intrinsic properties of the novae, they are of great importance for the comparison of the properties of our Galaxy with neighbouring systems like M31, M33, M81, M101. A thorough study of the novae in the Andromeda nebula, to extend the early results of Hubble's survey, is now under way with the 60-inch Mount Wilson reflector. It will provide full coverage for the two seasons 1953 and 1954, each extending from June to February.

A search for novae in the central directions of the Galaxy is being made by Haro with the Tonantzintla Schmidt telescope. It covers about 600 square degrees up to apparent photographic magnitude 13. Two plates a week are taken, and in addition one spectral plate (dispersion of 240 *A*/mm.) every month in order to detect faint novae showing H_α emission. Objects brighter than $m=10$ are observed photo-electrically, those between 10 and 13 photographically.

As was pointed out by Kukarkin, the combination of data from sky patrols carried out by observatories in different parts of the world can be extremely useful for obtaining well-covered light curves of novae.

A (2). OVERALL STRUCTURE: SPACE DISTRIBUTION AND MOTIONS IN THE DISK

The population of the disk consists of a great variety of objects with a wide range of concentration towards the galactic plane and of peculiar motions. The extreme population I objects like interstellar gas and dust and supergiant stars, show quite different properties with respect to space distribution and motions, compared to such stars as, for instance, the common G and K giants. The discussions at the conference have concentrated on the observational possibilities of studying the correlation—or the lack of correlation—between the space distributions of different kinds of objects. From recent theoretical as well as observational work it has become apparent that there is a wide range in the ages of the stars contributing to the disk population, from a few million so some 1000 million years. The differences in the distributions of the various kinds of stars very probably must be interpreted in terms of these differences in age and evolution. The most intriguing problem for the present thus seems to be this: Can we trace large-scale structure in the distribution of objects of different ages, and can we derive information on the evolution of the galactic system and of the stars themselves from the degree of resemblance in the large-scale structure exhibited by different objects?

Of great importance in this study is the comparison between our Galaxy and other stellar systems, particularly the nearest spiral nebulae. From these objects we derive information on the distribution of the most luminous stars of various types and of the interstellar medium and on their mutual relation. It is, therefore, appropriate first to summarize some of the results of the studies in this field communicated by Baade.

INFORMATION DERIVED FROM THE ANDROMEDA NEBULA

The Andromeda nebula is particularly suited for study of the properties of spiral arm and inter-arm populations, because it exhibits a well-defined spiral structure. Its arms are relatively thin and secondary branches which so easily confuse the picture are rare. Evidence of heavy absorption in the arms was derived from the fact that on H_α photographs emission nebulae appear by the hundreds in the arms—and the brightest and largest among them appeared also on the red plates used for the resolution of the general background, whereas none of them showed up on the earlier plates surveyed by Hubble. This indicated heavy selective absorption because the spectra of these emission nebulae differ in no way from those of the emission nebulosities of our own Galaxy.

On the other hand, the regions between the arms seem to be free from heavy absorption, as appears from the lack of reddening of globular clusters shining through them and from the observation of distant extragalactic nebulae through the inter-arm medium. Thus, the dust appears to concentrate in the arms. There are theoretical reasons to believe that dust and gas have the same distribution, and since in our Galaxy the spiral structure of the gas has been clearly demonstrated by the recent observations of the 21 cm. hydrogen emission line, we assume that in the Andromeda nebula too, the gas is concentrated with the dust in the spiral arms. Though, as was pointed out by Lindblad, there are some striking deviations of the lanes of dark matter of lower intensity in the inner region from the general spiral pattern.

The spiral arms of gas and dust are also the regions where the super-giants appear. The question rises whether gas and dust are primary and the stars secondary or whether the stars are primary and produce the gas and dust. We adopt the first alternative because:

(*a*) The short lifetimes of the super-giants require that these stars are continuously replenished and that star-formation is continuously going on in the dust and gas. There are reasons to believe that together with the super-giants, stars of all sorts of masses down to faint dwarfs are being formed in the associations.

(*b*) If gas and dust are the primary features of spiral structure, systems, otherwise similar to M31, but free of gas and dust, should be unable to develop spiral structure and no super-giants should appear. Such systems actually occur: the So nebulae, where the absence of gas may be due to collisions with other galaxies.

The stars observed in the central regions, the inter-arm region and the outermost regions, which are mainly population II objects (the resolution in these parts occurs simultaneously with the resolution of the globular clusters in M31) must be considered as disk population according to the shapes of the isophotes as measured, for example, by Hiltner and Williams. The disk population of type II seems to be the main source of light in the photometry of the nebula. Holmberg[1] found for the Andromeda nebula (NGC 224) the integrated colour index $CI = +0{\cdot}86$ which is very close to the mean value $+0{\cdot}88$ found by Stebbins and Whitford for the nearer elliptical nebulae.[1] The spiral arms with their population I stars are imbedded in this population II disk. However, it probably would be going too far to state that all the inter-arm stars are population II, because the spiral structure must have added new stars ever since the first disk stars were formed. Some of the newly formed stars will diffuse from the spiral arms into the inter-arm regions, and we should expect to find there a mixture of stars of all ages. The real composition of the spiral arm and inter-arm population in terms of populations I and II or in more refined classification into stars of various sub-systems or ages, is a problem that can be studied only in our own Galaxy.

Another result which is indicated by the study of the Andromeda nebula should be mentioned; from the surveys of the Cepheids and the OB super-giants it appears that both these objects occur in the spiral arms. But a high frequency of OB super-giants in a certain section of a spiral arm does not necessarily mean a concurrent high frequency of the Cepheids. What seems to be correlated is the frequency of the F, G and K super-giants and that of the Cepheids. For a final verdict three colour-measures are required to eliminate the ever-present reddening effects.

DISTRIBUTION AND MOTION OF INTERSTELLAR GAS

By far the most complete information on the distribution of the interstellar gas in our Galaxy is that being derived from the 21 cm. emission line of neutral hydrogen. It is based on measures of the Doppler shift, combined with an assumption concerning the rotational velocities in different parts of the Galaxy. The rotational velocities adopted up to now are based on the supposition that the mean motion of the gas is everywhere circular and shows rotational symmetry with respect to the nucleus. At distances from the centre equal to that of the Sun, the change of the velocity of rotation with the distance from the centre is immediately derived from the constants of galactic rotation and hence from proper motions and radial velocities of stars. At smaller distances from the centre this dependence can be derived, at least partly, from the radio measures themselves; for the outer parts of the Galaxy the radio measures will not provide such direct information and the interpretation of the measures depends largely on a model of our Galaxy. For recent results dealing with the distribution of the neutral interstellar hydrogen, we refer to an article by Van de Hulst, Muller and Oort.[2]

Measures of the continuous radiation at decimetre wave-lengths will furnish information on the distribution of ionized hydrogen.

Possible deviations of the mean motion of the gas from the direction of circular motion, and from rotational symmetry, can in general be detected only when mean motions and distances are measured for distant stars which are closely associated with the gas. Such are the super-giant stars and to the best present knowledge the Cepheids of population I. The discovery of very distant δ Cephei variables and OB stars, and the photometry and measurement of radial velocities of these objects is, therefore, of the greatest importance.

SUPER-GIANTS

Recent studies of the distribution of emission nebulae and of the O-associations have revealed the spiral pattern in the distribution of these objects up to about 3000 parsecs. An article by Morgan, Whitford and Code in the *Astrophysical Journal*, **118**, No. 2, 1953, gives the results described by W. W. Morgan at the meeting. Preliminary results of a survey of the southern emission nebulae by Bok, Bester and Wade were presented by Bok; this survey clearly shows the inner spiral arm running through the Carina concentration of early type stars. The distribution of the southern O-associations is being studied by Code from observations collected at the Radcliffe Observatory. The evidence available at present indicates that the spiral structure of the luminous O and B stars coincides with that exhibited by the neutral hydrogen.

We have no clear-cut information yet how the distribution of A and F type super-giants compares with that of the O-associations. In the nearby spirals the O, B and later type super-giants are intermingled in the spiral arms, but the percentage with which the various types are represented varies from place to place. Similar variations in our own system will have to be studied by means of accurate spectral classifications and photometry, using the finding lists of O–F super-giants published by the Tonantzintla, Warner and Swasey, and Yerkes Observatories.[1] The finding lists published up to now have a limiting magnitude of about 10, and extension to the 12th magnitude, which obviously is very important, is under way at the Tonantzintla Observatory. Particular emphasis, according to W. W. Morgan, should be laid on the homogeneity of the systems used for the spectral classification and the photometry, and it is strongly recommended that for the latter the UBV system,[2] applied in conjunction with the revised Yerkes system of spectral classification, be adopted. Further planning of these programmes was not considered desirable at the conference, as the observatories which are in a position to carry them out are already active in this field.

The problem of finding the OB stars fainter than the limits of the above-mentioned spectral survey was the subject of extensive discussions at the conference. Searching for very faint HII regions, as has been done with the 48-inch Schmidt, is successful in so far as it reveals those OB stars which are associated with the interstellar gas, but on account of the strong reddening it is often difficult to identify the exciting star (experience reported by Baade). Other methods which might be feasible, suggested by W. W. Morgan, are:

(*a*) Surveys of stars showing hydrogen emission lines, as carried out by Merrill and associates.[1] Their last catalogue, which lists early type stars showing H_α emission, is a suitable basis for the selection of some of the distant high luminosity OB stars. The objects listed belong to two categories of stars which happen to be well separated in luminosity, on the one hand low luminosity main sequence B0–B5 stars, with a range in absolute magnitude around -3 to -1 and, on the other, super-giants with luminosities around -6 or -7. From a plot of all the stars in this last list (which are rather fainter than those in the earlier lists) it appears that in certain ranges of galactic longitude the faint objects, which are about 11th magnitude, are strongly concentrated to the galactic equator. These very probably are distant and hence very luminous objects and their further observation might reveal a large fraction of super-giants. This method could in principle be extended to fainter stars. Experience at the Tonantzintla Observatory shows that emission stars up to the 14th magnitude can be found; actually, considerable numbers of new emission objects have been found already in a region centred at NGC 663.[2] The repetition of Merrill's survey up to $m=14$ is under consideration at Tonantzintla.

(*b*) Multi-colour photographic photometry might be used as a first step in sorting out the objects among which the luminous B-stars are to be found. As a next step accurate photo-electric multi-colour photometry could then pick out the objects sought. This procedure would rapidly converge and would be not too time-consuming.

That the criteria in searching for the O–F super-giants on the Schmidt objective prism plates at Cleveland and Tonantzintla have been successful in revealing a high percentage of real super-giants, has been confirmed by the observations with slit spectra. The situation is more difficult for types G2 and later. It turned out that the intensity of the CN band as a criterion for luminosity, which as a rule is a useful means of identifying the stars, is not reliable in all cases. No other reliable criterion has been found.

In addition to the photometry and accurate spectral classification of the stars found in the survey, determination of the radial velocities was strongly recommended. As has been stated already in connexion with the discussion of the 21 cm. radio measures, knowledge of the mean motions in different parts of the galactic system can be determined only from the radial velocities of distant objects, and the super-giants, with their low peculiar motions are best suited for this purpose. The problem of the determination of radial velocities for these faint objects is a difficult one, but the newly developed technique of objective prism radial velocities may well prove to be the solution (see also p. 41). The importance of such work can hardly be stressed too much.

CEPHEIDS

Cepheids can be identified at larger distances than O–F super-giants as no spectra are required. Therefore, extension of the existing surveys of Cepheids along the galactic equator to the faintest possible limits will be extremely valuable.[1] The survey should be supplemented with colour measures and, for the brighter objects, with observations of the radial velocities. It is particularly from the radial velocities of Cepheids that we may derive information about the velocity of rotation around the galactic centre as a function of the distance from the centre. The early work on radial velocities by Joy is now being extended to the southern hemisphere by Stibbs at the Radcliffe Observatory in Pretoria, but more remains to be done.

As had been pointed out, amongst others, by Kukarkin, the present surveys are still far from complete; for instance, new Cepheids have been found by him in some regions which hitherto seemed void of these objects. This was done in a survey of selected regions which reaches up to the 17th magnitude. The study of the distribution of the faint Cepheids is somewhat complicated by the fact that we have a mixture of variables of populations I and II, which can be separated by the shape of the light curves only if the periods are around 15–20 days, whereas separation is very difficult for periods of a few days only. However, for the studies of the most distant regions the large periods are obviously the most important ones.

In order to penetrate to very large distances, that is to reach the faintest Cepheids in the most transparent regions, the existing surveys (Harvard, Leiden, Sonnenberg) would have to be extended to fainter limiting magnitudes in the directions of low absorption. For the selection of the most transparent regions observations with the 48-inch Palomar Schmidt will be most valuable. Also recommended is a search for the brighter

Cepheids in the unexhausted areas for which many plates have accumulated already. Preliminary results from a survey of Cepheids in the Andromeda nebulae show that their distribution follows the spiral arms but is far from uniform. Similar conditions in the Galaxy are indicated by the crowding of Cepheids in particular regions like the Carina section. An important aim of the surveys must be to locate these rich distant concentrations.

The work on Cepheids appears to require much co-ordination, both in the planning of the faint survey and in the solution of the problem of measuring colour excesses. The latter is two-fold: how do we obtain intrinsic colours, and how will the colours of very faint objects be measured?

Fields to search for faint Cepheids

According to Baade, the selection of regions for faint extensions of the existing surveys will require very careful inspection of the 48-inch sky survey plates. Experience acquired in some provisional searching has shown that one gets occasionally, at a very low latitude, areas of remarkable transparency. A striking case is, for instance, the region around NGC 2158, a small distant cluster in the anti-centre region. Many extra-galactic nebulae appear here on 200-inch plates. Another field is the one of the strong Cygnus A radio source, where the total absorption could be measured from the colours of the brighter E nebulae; it amounts to 2·1 magnitudes. Such regions will be at least as useful in outlining the outer regions of the Galaxy as the general survey of the anti-centre region (Shapley). According to Oort, the selection of regions of low absorption may be facilitated in the future when the total amount of neutral hydrogen in different directions becomes more completely known from the radio measures.

As Morgan pointed out, it has to be realized from the beginning in these faint surveys that the directions in which the Cepheids will be found to be most numerous, will be correlated with low absorption. Still, this will not seriously hamper the study of the relation between the occurrence of Cepheids and, for instance, the early type super-giants, if directions of low absorption in the nearest arm are chosen for study of the conditions in the more distant arm. Additional information on the relation between the distribution of different kinds of high luminosity stars must, of course, be anticipated from the study of the Andromeda nebula.

The matter of selecting special regions for the study of faint Cepheids was delegated by the conference to a small committee, consisting of W. Baade, B. V. Kukarkin, P. Th. Oosterhoff, and the secretary of the continuation committee on co-ordination of galactic research. This committee might also act in stimulating and co-ordinating the further work on the variables in these selected fields.

Colours of the Cepheids

The determination of colour excesses of distant Cepheids is a very difficult problem and this probably is the reason why, apart from Eggen's programme,[1] very little has been done in this field until recently. The improvement of the photo-electric technique, however, will make possible great progress in this field in the future. A programme of photo-electric colours and magnitudes of southern Cepheids is being carried out at present by Walraven, A. B. Muller and Oosterhoff at the Leiden station in South Africa. The Cepheids are measured around maximum and the list contains all known objects brighter than $m = 14$ in maximum and south of declination $+30°$; part of the objects are followed over the whole light curve. Another programme was started at the Boyden station but it is doubtful if this will be continued. For the northern sky no large scale programmes have been announced yet. For the interpretation of the measures it will be necessary to have accurate knowledge of the normal colours of the Cepheids. This is probably the most intricate problem of all. There are too few Cepheids close enough to the Sun to be assumed free from absorption, and indirect methods will have to be applied. The most promising one might be the application of multi-colour photometry to the foreground stars in a small field around the nearer objects, in a way outlined by Strömgren (see p. 28). This might result in a three-dimensional charting both of the stars between the Sun and the Cepheids and of the distribution of the absorbing medium. The subject requires more experimental work and its organization may be a task for the continuation committee mentioned on p. 2. Discrimination of population I and II Cepheids is an important point in all these investigations.

LONG-PERIOD VARIABLES

Another important class of variables, and one which deserves much more attention than it has received in the past, is the long-period variables. On the whole, these objects form a class intermediate between the more typical population I and II stars and they can be studied up to large distances. The dependence of their kinematical and distributional properties on the period makes them particularly interesting,[2] and their large numbers allow rather detailed studies of this dependence.

For future work, the emphasis should be in the first place on homogeneity in the discoveries. In some of the large surveys of variable stars in the past, few of the long-period variables have been discovered either because only pairs of plates with short time intervals were blinked or because the plates taken were not sufficiently uniformly distributed over a period of at least a few years to allow the determination of the period. Additional observational work may, therefore, be required. In view of the rather simple procedure of determining periods and light curves, the work on these variables would soon give valuable results.

The new survey should be combined with radial velocity and proper motion measurements for the brighter objects (see Section C).

SURVEY OF OTHER VARIABLES IN THE DISK

Interesting results on a survey of T Tauri stars were announced by Haro, indicating a correlation between their spectral characteristics and the physical conditions in the interstellar medium. These results have meanwhile been published in the Tonantzintla and Tacubaya *Bulletin*, No. 8. Similar work for fainter stars in small regions of the dark cloud complexes is being done by Herbig at Lick.[1] This work does not require further co-ordination.

OPEN CLUSTERS

The modern view on stellar evolution has a very interesting bearing on the study of open clusters. Information on the age of the clusters may be obtained from the study of their physical characteristics, and this should be combined with the study of their space distribution. As was stressed in particular by Ambartsumian, we know that the clusters which contain O-type stars are associated with the O associations and hence follow the spiral pattern of the OB stars. We would like to know the distribution of clusters in which the brightest stars are of types B or even A, and which sometimes contain late-type giants. If these are clusters which some time in the past contained O stars, we want to compare their distribution with that of the late-type giants in general in the Galaxy. More generally, from the comparison of the distribution of clusters of different physical characteristics we may derive either information concerning the evolution of the stars or concerning the evolution of the spiral pattern in the Galaxy.

The work on clusters can be divided into the search for fainter, distant clusters not listed up to now, the photometric and spectroscopic study of as many objects as is possible, and the measurement of radial velocities.

The existing surveys by Shapley, Trumpler, and Collinder contained clusters in which the brightest stars are brighter than the 16th magnitude. Undoubtedly the Palomar sky survey will increase the number of known open clusters.

For the distance determination of the faintest objects, where photoelectric methods would be too time consuming, the photographic three-colour method developed by W. Becker has given very promising results.[1] As was shown by Becker, the distribution of clusters can be determined in this way up to distances including the spiral arms next to the one in which the Sun is located, and a close coincidence of the seventeen clusters studied with the arms was found. If a sufficient number of stars are measured in the cluster the precision of the distance determination depends mainly on the accuracy of the zero point of the photometric scale; an accuracy of 10% in the distances had been reached in the cases shown.

Becker intends to extend this work to the southern hemisphere and to some other regions of particular interest. A large scale programme all along the galactic circle would seem very valuable. According to the experience reported by Heckmann, Schmidt cameras of moderate sizes will be most useful, with focal lengths about 1½ m. and apertures about 40–60 cm.; they can reach a limiting magnitude of about 19. This might well be a very attractive programme for small observatories. The photometric system to be used requires careful consideration. The necessity of having reliable standards well spread along the Milky Way was stressed at the conference; the details about the photometry were not discussed. The matter of choosing standard sequences is, of course, to be considered in conjunction with the same problem in variable star work.

It was felt by the participants at the conference that it would be of great value if the important observations on open clusters accumulated by Dr Trumpler at the Lick Observatory during the past 20 years would become available to astronomers working in this field. The following is derived from a letter by Dr Trumpler of May 1953.

1. Unpublished material. Radial velocities of stars in 74 galactic star clusters divided as follows:

187 stars in the Pleiades, Praesepe, and Coma;
234 stars in 16 clusters nearer than 800 parsecs;
400 stars in 55 clusters beyond 800 parsecs.

The observations and measurements have been completed. Dr Trumpler is now occupied with a discussion of systematic errors and a reduction of the observations to a more homogeneous system based on observations of numerous bright standard stars. The spectral types of the stars have also been classified from the slit spectrograms. Spectral classifications from slitless spectrograms are also available for fainter stars and in other clusters, not observed for radial velocity.

2. The above mentioned data will be used for the study of galactic rotation. For this purpose distances will be re-determined using magnitudes, spectral types, and colour indices.

3. The radial velocity data will also be used for a number of subsidiary problems (spectroscopic binaries, relativity red shift, super-giants and peculiar stars in clusters, internal motions, and peculiar velocities of the clusters).

4. It is urged by Dr Trumpler that radial velocity observations of galactic star clusters south of $-30°$ declination be contemplated.

The conference expressed the hope that these extensive observations will soon be published.

INFRA-RED SURVEYS OF M, N, AND S STARS

From the report given by Dr Nassau we summarize the following points which are of interest in the discussion of the population of the galactic disk. The work being done at the Warner and Swasey Observatory deals mainly with:

(*a*) The M-type stars, M5 and later, characterized by the presence of TiO bands at λ 7054, at the telluric A band, and at λ 8432. Among these stars a number show the VO bands and are variable. (See *Ap.J.* **119**, 175, 1954);

(*b*) The N stars characterized by strong CN bands. (A list of these is published in *Ap.J.* **120**, 129, 1954);

(*c*) The S stars recognized by the presence of LaO bands, and

(*d*) The early M suspected super-giants, recognized by extremely high reddening and their wedge-shaped appearance on the spectral plates; these latter stars were picked out in the present survey when they showed stronger reddening than the super-giants of this type in *h* and χ Persei.

The first group, late M stars, shows a uniformly varying density, decreasing with increasing distance from the galactic centre. The VO stars among them appear to show the same distribution as the group as a whole. Reference to the study of these stars in connexion with the density distribution in the nuclear region was already made on p. 12.

The second group, the N stars, appears to be of particular interest. Their distribution in the Galaxy is quite different; they show clustering, seem to be associated with obscurations, and show stronger galactic concentration than the late M giants. The number of N stars on which the study is based is over 300. The counts in different directions in the galactic plane show larger numbers along the spiral arms of interstellar matter and OB stars, than outside these. Very few N stars were found in the directions of the Scutum and Sagittarius Clouds where the late M stars are very abundant. Although they show a definite tendency to occur in groups, they do not seem to be associated with the clusterings of OB stars. They can be found up to very large distances, of the order of the dimensions of the galactic system. Further study of these stars may give very valuable results. Very little is known about their absolute magnitudes. It was suggested that one way of determining these would be to measure the intensity of the interstellar D lines in directions where these intensities can be calibrated by means of early-type stars. A more direct determination would be from the proper motions and radial velocities. A list of the brighter objects of this class is being compiled and will be suggested for meridian observations.

In some cases the S stars appear to be associated with the groupings of OB stars but it is too early to state that they belong to the same aggregates. In a number of cases the S stars appear in groups. A list of 31 new S stars is in press.

The suspected early M super-giants show high galactic concentration, they appear in groups and in some cases in O associations. A list of 89 is to be published in 1954.

The close association of the super-giant M stars with the associations of OB stars (h and χ Persei, the probable association of α Orionis with the Orion group) presents an evolutionary problem of great interest and should stimulate the further surveying of these stars.

Infra-red surveys of the type described by Nassau are being extended to the southern hemisphere at the Harvard Boyden Station[1]; this work is planned in the same way as the Cleveland Survey (covering a belt of 4° width along the galactic equator). The Magellanic Clouds will also be studied.

B (1). LOCAL STRUCTURE: DISTRIBUTION OF DIFFERENT TYPES OF STARS IN THE PLANE

Numerous extensive studies of the space distribution of stars of different spectral types, based on star counts and colour measures, have in the past been published for different sections of the Milky Way.[1] These studies have given information on the nearby obscuring clouds, and they have revealed marked differences between the distribution of different kinds of stars. They have not, however, given us the much desired insight into the large-scale features of the density distribution.

With the new developments in our knowledge of spiral structure of the interstellar gas, and the increased information to be expected from the planned observations of super-giants described in section A (2), the fundamental question with regard to the normal giants and main sequence stars can be formulated as follows. We have theoretical reasons for believing that as we go down the main sequence from the OB super-giants to the stars of lower luminosities, the main sequence B stars, the A stars and F stars, we are considering stars in the order of their ages. Until we come to the G main sequence stars, we very probably are dealing with stars which are younger than 3×10^9 years. The later type giants may also be included among these. Can we, going up these steps in age, trace differences in the large-scale structure in their distributions which might give information on the distributions at the times of formation?

The lower luminosity of these stars with respect to the super-giants limits the investigation to a much smaller region around the Sun. When we consider stars like main sequence A stars or ordinary K giants, that is of absolute magnitude around zero, and if we assume 2·5 magnitudes absorption, the distance up to which we may hope to get reasonably accurate information is about 1500 parsecs, corresponding to a limiting apparent magnitude about 13·5. This, however, reaches far enough to allow comparison with the section of the spiral arm of gas and OB super-giants in which the Sun is located. One of the first things to be done will be to determine the density distribution of, for instance, the A stars, in directions along and across the direction of the spiral arm. As the shortest distance to the next arm is 2000 parsecs, an investigation up to 1500 parsecs would penetrate far enough into the inter-arm population.

From the observations in the Andromeda nebula we know that the major constituent among the inter-arm giant stars is the population II stars which are resolved simultaneously with the same kind of objects in globular clusters. This probably is also the case in the Galaxy, the population II giants forming a continuous medium in which the population I stars are imbedded. It is, therefore, of primary importance that in the investigation indicated above there should be no confusion between the giants of the two population types. We should select only those objects for which reliable criteria to discriminate are available.

SELECTION OF TYPES OF STARS FOR STUDY

The stars to be picked out from surveys for further investigation should meet the following requirements.

(1) They must represent a narrow region of the HR diagram, that is, they must be defined by narrow limits of absolute magnitude and intrinsic colour.

(2) It should be possible to pick them out up to distances of the order of 1500 parsecs.

(3) It should be possible to distinguish between population I and II objects.

(4) It must be possible to pick out identical groups of stars among the very faintest objects (about $m = 13$ or 14) and among the bright ones ($m < 8$), in order to make the comparison of nearer and distant regions possible.

From the discussions at the conference it became clear that there are only few groups which meet all these requirements, but these offer an interesting choice.

(*a*) The most promising group is that of the main sequence A0 stars. From Morgan's description of the natural groups in spectral classification,[1] as well as from the experience of W. Becker, it appears that these stars can be picked out on objective prism plates of very low dispersion (1000–2000 A./mm.), being recognized by the maximum strength of the H lines. It was stressed by different participants, that it will be necessary to limit the group to the A0 stars only and not to take the B9–A3 stars, in order that the four requirements just mentioned will be best satisfied. The main sequence A0 stars can even be found up to $m = 14$ without real difficulty. The best procedure might be to search for all the objects with maximum and near maximum intensity of the Balmer lines, and next to narrow down the selection of those with strongest lines by photometric tracing of the spectra. Tracing might also help to guarantee the uniformity of the selection of bright and faint surveys. Some widening of the spectra is recommended to facilitate the first visual step.

As was pointed out by Bok during the conference, earlier and current investigations of the A stars reveal a density distribution which does not seem to resemble that of the OB stars. For instance, an increase of the density in the direction to Carina at distances of 500–1000 parsecs is found, although this region is located between the spiral arms of the OB stars. In the directions of Sagittarius, Cygnus, and Monoceros, a drop in the A star density is indicated, according to Bok. The deviations in the motions of the A-type stars with respect to the later types (solar motion, distribution of peculiar motions) also reflect irregularities in the space distribution of these stars. These facts, combined with the presumption that these stars are older than the OB super-giants but younger than the majority of the common stars, make them very interesting objects for the proposed survey.

(*b*) Next in importance in the study of local structure seem to be the F2–F8 weak-line and strong-line stars. From Miss Roman's studies[1] of the differences in the kinematical properties of these stars it is very probable that they represent different stages of stellar evolution, the strong-line stars possibly being close to the A stars in kinematical and distributional properties, whereas the weak-line stars rather have to be considered as an intermediate stage between populations I and II.

The differences between these two categories and the problem of their relation with other types of stars make them interesting objects for the present studies. The two groups can be distinguished up to $m = 11{\cdot}5$ on objective prism plates (dispersion 230 A./mm. between H_γ and H_δ) of a good quality, and perhaps even fainter if special effort is made. The objective prism survey must be followed by slit spectra for a more refined classification.[2]

(*c*) A third group which may be well suited for this investigation is the natural group of K-giants. The main problem is here, the segregation of population I and II objects. The best way to deal with these stars might be, to pick them out from objective prism surveys without the segregation of the various types and have this latter done by individual spectroscopic observation afterwards. The survey of the objective prism spectra could probably be based on the CN break at λ 3863. The observation of some hundreds of stars up to $m = 14$ for slit spectra with a dispersion of 150 A./mm. will require a powerful instrument, but the importance of the studies would justify the effort. It is not impossible that one of the largest telescopes may become available for this work. According to Baade, six-colour photometry as was employed by Stebbins and Whitford might also be able to distinguish between the two types of population. This would require some more experimental work.

FURTHER WORK ON THE INDIVIDUAL STARS FOUND IN THE SURVEY

The determination of the colours and absolute magnitudes of the selected stars must be based on accurate photometry at different wave-lengths. The multi-colour method now being developed by B. Strömgren may provide accurate luminosity and spectral classifications for the types of stars considered, except perhaps for Ao.

This method is based on photo-electric measurements of the intensity in narrow wave-length regions. For the B, A and F stars one uses the intensity of H_β and the Balmer discontinuity. For the K stars the strength of the K line, the CN band and the G band. Preliminary results indicate a very high accuracy for the absolute magnitudes and colours measured in this way for the F stars, and a somewhat lower accuracy, but still better than that found by other methods now available, for the B and K stars. The separation of population types I and II among the K stars remains a difficult problem. For the early A stars the method is less accurate.

A most important aspect of the multi-colour method is that it allows measurements of fainter stars than spectroscopy can reach with an equal observing time; an accuracy of 0·2 magnitudes in the absolute magnitude, and of 0·03 in the spectral class can be reached for F stars which are about two magnitudes fainter in the multi-colour method than in the direct spectroscopic observations. The method does allow detection of some peculiar stars like metallic line stars, for which the predicted luminosities are somewhat lower than for normal stars of the same Balmer discontinuity.[1]

In the case of the A-type stars it is possible that the selection by the method of the last section is narrow enough to ensure a small range in colour and luminosity. This, however, would have to be tested by means of samples. It will hardly be possible to investigate all the stars found in the surveys. The solution might be to observe representative samples of stars of different apparent magnitudes, large enough to allow conclusions with regard to the distance distribution.

CHOICE OF THE FIELDS

It was the general opinion of the participants at the conference that the above-mentioned programmes require preliminary work to find out if the approach outlined above is feasible. It is, therefore, recommended that one should start with a very limited programme containing two or three test areas in the Milky Way and chosen so that they will give information on the density distribution along and across the spiral arm. The following areas were recommended:

(*a*) An area at $l=45°$ (Cygnus Cloud), i.e. in the low latitude field number 3 studied by Nassau and MacRae; this field lies in the direction of the spiral arm passing through the Sun.[1]

(*b*) An area at $l=92°$, centred on Kapteyn S.A. 8. In this area we would have proper motions available which may prove to be an important aid in analysing the photometric and spectroscopic data. The absorption seems to be low. The distance to the next spiral arm in this direction is about 2500 parsecs. An alternative choice suggested at the conference was an area at $l=70°$, $b=-2°$, which seems to be unusually transparent and has already been investigated by several authors. It would, however, be a disadvantage that here the inclination to our spiral arm is only 40° and hence the survey might not reach sufficiently far outside the arm. Another choice would be $l=97°$, an area for which spectral plates are already available at the Warner and Swasey Observatory and which also seems rather transparent. Here, however, there are no data yet available on the proper motions.

(*c*) Possibly an area at $l=175°$ or at $l=205°$. In the direction $l=175°$ our spiral arm shows branching and one probably would not reach outside the arm within the first 1500 parsecs. The investigation might show to what extent the somewhat irregular distribution of the OB stars is also present among the other types. According to Bok, the obscuration is low around $l=175°$, near NGC 2244. At $l=205°$, radio measurements (21 cm.) indicate low interstellar gas density up to 1 kiloparsec, next, a high maximum at 1·4 kiloparsecs, and again very low density between 2·0 and 3·0 kiloparsecs. The density distribution of the gas is less complicated than at $l=175°$.

The final choice between these fields will be a matter for the continuation committee to decide.

Dr Nassau thought it might be quite possible for the Warner and Swasey Observatory to take the objective prism plates which are required for the first steps in the surveys mentioned above.

SURVEY OF AO-TYPE STARS

In the case of the Ao-type stars it would be interesting to add to the above programme a survey of all main sequence Ao-type stars occurring on the plates taken for the survey of the super-giants ($m < 12$). From this, one might obtain a complete picture of the distribution of these objects within 1000 parsecs from the Sun.

SURVEY OF WEAK-LINE AND STRONG-LINE STARS

Different participants at the conference were of the opinion that an extension of Miss Roman's survey of weak-line and strong-line stars beyond $m = 5{\cdot}5$ would be very important. This would allow a more detailed study of the kinematical properties of the two groups. It would be especially interesting to know, for instance, what the distribution of the peculiar motions is for the group of the strong-line stars, and how closely it resembles the distribution found for the A-type stars. In this way it might be possible to indicate to which age-group these stars must be assigned.

Such work might also throw more light on the question of the continuity or discontinuity of the physical and kinematical properties of the main sequence stars. As reported by Parenago (see the translation of his speech in the *Astronomical Newsletter*), Einesto has found evidence of the composite nature of the main sequence F5–G9 stars from the velocity distribution of these objects, supporting Parenago's and Masevich's results on the existence of two overlapping main sequence series.

B (2). LOCAL STRUCTURE: WORK IN HIGH LATITUDES

The potential possibilities of research in Kapteyn's Selected Areas at intermediate and high latitudes, where magnitudes and proper motions are already available, could be fully exploited if a more accurate spectral and luminosity classification would become available, especially for the later type stars. The purpose of work on faint stars in these latitudes is manifold. At the highest latitudes the improved data can serve for a new determination of the density distribution and of the force perpendicular to the galactic plane as a function of the distance z to the plane, $K(z)$. The limiting magnitude may be set here at $m = 13{\cdot}0$ (photographic). At intermediate latitudes one would hope to find the correlation between the density at some distance above the galactic plane with the density in the plane. Here the limit should be set at 13·5 or 14·0, so that G and K giants can be reached up to distances of 2 to 3 kparsecs from the Sun.

Another important problem is the velocity distribution parallel to the galactic plane as a function of the distance z. For this study we need accurate photometric distances and proper motions. Problems like the deviation of the vertex as a function of z or, more generally, the detailed study of the velocity distribution for stars of different populations might in this way be attacked.

The spectral classifications may also provide more information on the physical characteristics of the stars at large distances from the galactic plane. The necessary luminosity determinations can be based, perhaps, on accurate multi-colour photometry as described in the preceding sections.

Work at high latitudes was reported by various participants at the conference. Parenago mentioned work by Artuchina on the proper motions of 1000 stars in a region of 25 square degrees up to the 15th magnitude, indicating that it might be possible to determine the percentage of stars belonging to various sub-systems by studying the velocity distribution at different distances from the galactic plane.[1] This work, however, also showed the need for accurate proper motions for much fainter stars. Münch, at Mount Wilson Observatory, in collaboration with Haro, is observing spectra of the stars in Malmquist's catalogue of colours and magnitudes in a field of 100 square degrees. Spectra are being obtained with the 200-inch telescope for the stars which are picked out as G and K giants.

Harris, at the Yerkes Observatory, has measured photo-electric colours with the 82-inch McDonald telescope for stars in an area of 35 square degrees. Classifications in this region have been made by Nassau. Münch has also observed spectra of the blue stars near the galactic pole found by Humason and Zwicky and found that a large percentage of the bluest stars does not show H lines and has at the same time other abnormalities. This survey may be the basis for further investigations of these stars.

Most of these investigations have not, however, a direct bearing on the problems described above in connexion with the stars' motions. For these and also for the study of the general density distribution, the work should undoubtedly be concentrated on the Kapteyn Selected Areas. It was decided at the conference that the co-ordination of efforts for various types of measurements in the high and intermediate latitude areas will have to be organized by the continuation committee. It will have to decide first on the choice of the Selected Areas, and next consider and co-ordinate the proper motion work (which possibly will have to be organized by the commission for the Astrographic Catalogue), the radial velocity work (possibly by Fehrenbach's method), multi-colour photometry and spectral classification. In this connexion it is important to notice that a survey of slit spectra for stars brighter than photographic magnitude 12·0 in some high latitude Selected Areas has been undertaken by Miss Roman with the 82-inch McDonald telescope.

C. PROPER MOTIONS AND RADIAL VELOCITIES

One of the most important fields for co-ordination in galactic research is the organization of programmes for proper motions and radial velocities. The need for accurate and more extensive data on stellar motions is strongly felt in connexion with the studies of the structure of the Galaxy and of the relation between the physical and the kinematical characteristics of the stars. Not long ago it seemed that meridian astronomy and photographic astrometry were rather remote from the main problems in astrophysics. It is realized now, that the study of stellar motions gives highly important information on the evolution of the stars and of the stellar system. A detailed account of the needs in this field, as discussed at the conference, follows:

(I) PROPER MOTIONS: PHOTOGRAPHIC ASTROMETRY OF FAINT STARS

Variable stars

Among the variable stars we find the most pronounced types of halo stars and typical disk population II. They are the main source of information on the motions of population II objects. They are scarce objects in the neighbourhood of the Sun, and their study, therefore, requires observations up to $m_{pg} = 12$ or fainter.

Current programmes at various observatories are:

The determination of proper motions of RR Lyrae stars at the Mount Wilson Observatory, in collaboration with the Leiden Observatory, and at the Leander McCormick Observatory.

The determination of proper motions of long-period variables at the Leander McCormick Observatory and at the Leiden Observatory.

The determination of proper motions of faint variable stars by different astronomers at the Sternberg State Astronomical Institute, Moscow, by comparison of plates taken with the 38 cm. astrograph with first epoch Astrographic Catalogue plates, or with first epoch plates taken at Moscow since 1932. (New results for 36 RR Lyrae stars have been published recently by Pavlov.[1])

Although these programmes will soon considerably increase our knowledge of the motions of these objects, any attempt at further work on them should be encouraged. It is a favourable circumstance that some of the observatories which took part in the early work on the Astrographic Catalogue have indicated a desire to use the old epoch plates for proper motion work. Therefore, a list has been prepared by Dr L. Plaut of the Kapteyn Laboratory, of variable stars brighter than photographic or photovisual magnitude 12·0 in maximum, occurring in Parenago and Kukarkin's catalogue. The list contains 2000 RR Lyrae, long-period, semi-regular and RV Tauri variables, and these are arranged in the order of the Astrographic Catalogue zone in which they occur. δ Cephei variables are not included; because of the smallness of their proper motions, these can be studied only with meridian instruments. The list will be made available to all institutions interested in this work, through the intermediary of the Commission for the Astrographic Catalogue, and may be published in the next report of that Commission. Preliminary work in this field shows that about half of the stars in this list do indeed occur in the Astrographic Catalogue. Among the missing ones, many can be measured on Astrographic Chart plates when these are available.

This list may be followed by others when the need arises for measurement of other types of stars of special interest.

For almost all purposes for which proper motions are studied, it is necessary that the relative proper motions are reduced to an absolute system. This is always necessary, for instance, when proper motions are converted into linear space velocities. One way of doing this, is to assume a mean proper motion of the reference stars by adopting the most probable value for the parallactic motion and for the differential galactic rotation. A more satisfactory procedure is that of reducing the motions to a fundamental system. (See also Vyssotsky's discussion in the Reports on the Dearborn Astrometric Conference, *A.J.* **59**, No. 2.) Reduction to a fundamental system will be possible for stars of the magnitude range measured on the Astrographic Catalogue plates (photographic magnitudes 9–12) when a repetition of the AGK2 (see Section C (3)) has been carried out.

In view of these considerations Parenago proposed that, together with the results for the stars of the particular programme, the reference stars, their apparent magnitudes and their measured relative proper motions are also published. One might even consider measuring the AG stars present on the plates in order to make sure that reduction to an absolute system can be accomplished as soon as the AGK2 repetition is completed.

It must also be emphasized here, that in those cases where the reduction to absolute has to be based on the parallactic motion of the reference stars, at least twenty reference stars are needed to reduce the influence of their peculiar motions to a satisfactory minimum.

Kapteyn's Selected Areas

Proper motions in the northern Selected Areas have been published by the Pulkovo and Radcliffe Observatories.[1] The southern areas are now being measured at the Kapteyn Laboratory on plates taken at the southern station of the Yale Observatory.

Repetition of these Radcliffe and Pulkovo observations in the northern Selected Areas could give a ten-fold increase of the weight of the proper motions, and probable errors of the order of $0''.001$. Such very accurate data would allow a much more refined study of the velocity distribution of distant objects than was hitherto possible. The great amount of work involved might make it desirable to confine the repetition to part of the areas. In that case, preference should be given to the low galactic latitudes (for instance, those below 20°) where the proper motions are smallest and where the motions are most complicated. These low latitude areas cover $40' \times 40'$ and the limiting photographic magnitude is about 14·5.

Even if such a programme is undertaken, there will remain a strong need for proper motions and radial velocities of many more low latitude stars brighter than 12·5. This is the limiting magnitude up to which we eventually may expect to get accurate spectral classifications and radial velocities. It is also the limiting magnitude of the Astrographic Catalogue. Therefore it was stressed, particularly by Oort and Van Rhijn, that a very important task for the observatories equipped with Astrographic Catalogue refractors will be the determination of proper motions in regions centred on the Kapteyn Selected Areas, but much larger than those given in the Pulkovo and Radcliffe Catalogues. The choice of the Kapteyn Areas is obvious, because here we have already magnitudes, colours, and spectral classifications available. The size of the areas should be taken $3·5° \times 3·5°$, as in the Bergedorfer Spectral Durchmusterung. (The Potsdam Spectral Durchmusterung, which covers the southern Kapteyn Areas, has even larger fields and reaches stars of about 12·0 and brighter.)

An average low latitude area of $3·5° \times 3·5°$ contains, between photographic magnitude 11·5 and 12·5, about 150 A stars, 250 G stars and 60 K stars. With the subdivisions into giants and dwarfs and possibly also into strong- and weak-line stars, there remain only relatively small numbers of stars per area for statistical investigation. It is, therefore, proposed that the project includes all Kapteyn Selected Areas below 10° galactic latitude. A list of these areas and their co-ordinates follows.

Selected Areas below 10° galactic latitude

No.	R.A. (1900)	Decl. (1900)	l	b	No.	R.A. (1900)	Decl. (1900)	l	b
8	1^h 0^m	+60° 10′	92°	−2°	110	18^h 37^m	0° 00′	0°	1°
9	3 4	60 20	106	3	123	7 65	−15 00	197	0
18	21 24	60 10	68	6	134	18 10	−15 00	343	−1
19	23 23	60 00	81	−1	147	7 3	−30 10	209	−9
23	3 39	45 00	120	−7	148	7 58	−30 10	215	1
24	4 39	44 50	128	0	157	17 25	−30 10	325	1
25	5 37	44 50	133	9	172	8 40	−45 10	232	−1
39	19 47	44 50	47	9	173	9 35	−45 10	239	6
40	20 47	45 00	53	0	179	15 48	−45 10	301	5
41	21 50	45 00	61	−8	180	16 46	−45 00	308	−2
49	5 24	29 40	145	−2	192	9 26	−60 00	248	−7
64	19 58	30 00	35	−1	193	11 27	−60 10	261	1
74	6 15	15 10	163	1	194	12 57	−60 10	272	2
87	19 11	15 00	17	0	195	14 57	−60 10	286	−3
98	6 47	− 0 10	181	1					

In addition, this programme of proper motions and radial velocities should include the high latitude areas mentioned in Section B (2), which will be selected by the continuation committee.

Future matters for organization will be the extension to some other particularly transparent low latitude regions and to some areas at intermediate latitude.

Another suggestion for proper motion measurements are the remaining regions in which Fehrenbach has measured objective prism radial velocities (see p. 41).

Milky Way regions of special interest

Proper motions in the low latitude fields, in which luminosity classifications have been made at Cleveland, would be of great value for the statistical study of these data. (See reference no. 1 for page 29.) The same holds for the test areas mentioned in Section B (1).

T Tauri stars

Surveys of T Tauri stars are now being carried out at the Tonantzintla and Lick Observatories (Haro, Herbig). Work on proper motions of the nearest objects (Parenago, Pels and Uranova) indicates that their motions are fairly large (± 12 km./sec.) and more precise measurements especially of the nearer groups of these stars within 500 parsecs, may prove very valuable in the interpretation of these objects. (See also p. 44.)

Stars in the outer regions of open clusters

For most clusters, only the central part has been thoroughly studied. Measurements in the outer parts will be important in order to determine how far from the centre the cluster extends. A survey of the Hyades in a region 5×5 degrees up to photographic magnitudes 14·5 and in a region 10×10 degrees around Praesepe, $m < 10$, is reported by Heckmann. A search for fainter members ($m < 12$) in a large area around the Hyades (about 30×30 degrees) is being carried out by Pels at the Leiden Observatory. It is based on a comparison of Astrographic Catalogue positions with modern ones. Work done at Hamburg indicates expansion in the Eddington Perseus cluster. It would be desirable to study this also in other clusters.

Late M-type giants found in the infra-red survey at Cleveland

Proper motions of these stars will help in the determination of their absolute magnitudes, about which there is considerable uncertainty at present. A list of 120 of the brighter objects (photographic magnitudes < 13) has been made available by Dr Nassau and will be sent to institutions interested in this work.

Stars with large proper motion

The study of the faint stars with large proper motion is important for the determination of the faint end of the luminosity law and also because among these we encounter dwarf stars with high space velocities. Some of these have come from the inner regions of the Galaxy and are, therefore, particularly interesting for further spectroscopic investigation.

For both these problems we require a more accurate determination of the proper motion than is available at present. A programme of such measures could very well be carried out with the aid of the Astrographic Catalogue or chart plates. The stars to be measured could be picked out principally from the lists by Luyten, Ross and Wolf.

Luyten's Bruce Proper Motion Survey covers the southern hemisphere. A list of the stars with large proper motions found in this survey (most of them exceeding 0″.05) for declinations below $-50°$ has been published in the General Catalogue of the Bruce Proper Motion Survey. It covers all stars up to about photographic magnitude 14·5 and many fainter ones.

Luyten has published a separate list of all stars with proper motions in excess of 0″.5 and negative declinations (904 stars) and other lists containing 1054 stars with proper motions between 0″.5 and 0″.3 and declinations south of $-40°$, with approximate apparent magnitudes and colours.[1]

These lists are essentially complete. Another similar list[1] of 393 stars within the same limits of proper motions and between declination 0° and −40° is incomplete. Further, a list of 305 stars north of the equator with proper motions between 0″.2 and 0″.5 has been published by Luyten.[2]

For the northern hemisphere, Ross' lists[3] are the most extensive ones. They contain stars brighter than the 13th photographic magnitude in about 85 % of this hemisphere (and 20 % of the southern one). Ross lists about 800 stars with proper motions larger than 0″.3. Wolf's lists[4] contain about 1500 stars, mostly brighter than $m=13$, about one-third of which have proper motions exceeding 0″.3.

The limit of 0″.3 would be a suitable one for the selection of stars for re-measurement. If the still unpublished lists compiled by Luyten would become available the total number of these stars would be about 4000, about 1500 of them being bright enough for the Astrographic Catalogue plates and many of the remaining ones measurable on the chart plates.

Part of these stars are covered by other independent programmes. For instance, the Ross stars will be measured on 40″ plates taken at the Yerkes Observatory. There is no reason, however, to avoid duplication.

Of special interest will be the improvement of the proper motions of all the white dwarfs known at present. A study of the colours and magnitudes of white dwarfs is being made by Harris at the Yerkes Observatory. Various lists of white dwarfs have been published by Luyten.[5]

Stars in the O associations

For this programme see Section D (1), pp. 43 and 44.

(2) PROPER MOTIONS: MERIDIAN OBSERVATIONS

There are several fundamental problems with regard to galactic structure, which require precise knowledge of proper motions of bright stars ($m<8$) in a well determined fundamental system. Examples are: the determination of the constants A and B of galactic rotation; the determination of the zero point of the period luminosity curve of Cepheids; the calibration of spectroscopic absolute magnitudes of high luminosity stars. It was decided by the conference that a list of stars for meridian observations be suggested to meridian observers. A first list, containing stars north of declination −20°, has been made and is added as an Appendix to the present report.[6] It contains the following stars:

Cepheids (period longer than 1·5 days), brighter than 8·0 visual magnitude in maximum.

O–B5 stars classified by W. W. Morgan in the revised system of the Yerkes Atlas, brighter than 8·5 visual magnitude.

O–B3 stars according to the Henry Draper Catalogue, still unclassified in the Yerkes system, also brighter than 8·5.

OB super-giants brighter than 8·5 visual magnitude, found in the survey by Nassau and Morgan.

The list contains 1061 stars. Of these, 274 marked 'P', may be given priority by those institutions who want to work only on a limited programme; they include all the Cepheids, the super-giants with either a distance modulus below 10·0 or a visual apparent magnitude (corrected for interstellar absorption) brighter than 6·0, and a few additional stars in the nearest associations. For each star are given the BD number and the approximate apparent visual magnitude. Some of the listed stars are among the fundamental stars regularly observed (FK3 stars, etc.). A few observatories have already indicated that they may carry out at least part of these observations. It is hoped that others will follow so that the highest possible accuracy for these objects can be reached.

For Cepheids, a table of the brighter objects with a list of early epoch references has been published by Parenago in *Variable Stars*, **6**, 102, 1947. Parenago stated that a card catalogue of meridian and astrographic positions of variable stars is kept at the Sternberg Institute; the information on these cards will be made available upon request.

It is intended to publish other lists of stars of particular interest for meridian observers as the need arises. The M super-giants and the N and S stars referred to in Section A (2), may be among these.

(3) PROPER MOTIONS: THE REPETITION OF THE AGK2

As was reported by Heckmann, the Hamburg Observatory contemplates repetition of the AGK2 between the years 1956 and 1960. This large project consists of two parts: (*a*) the photographic repetition of the 1930 plates and measurement of the relative proper motions; and (*b*) the meridian observations of reference stars during the same years. Details concerning this project are published elsewhere.[1] An important question discussed at the conference was, whether the repetition should be made about 1970–80 as planned originally, or already about 1960 with the consequent reduction in the accuracy of the proper motions obtained from the comparison with AGK2. The repetition would produce a homogeneous system of 'absolute' proper motions for about 180,000 stars down to the 11th photographic magnitude in the northern hemisphere. Their mean errors would be $\pm 0''.008$, for the 1960 repetition.

The importance of this project lies not only in the large number of proper motions of faint stars which will become available for study of various galactic problems, but particularly in the fact that it will furnish the bases for both a more accurate reduction of the old meridian catalogues (magnitude errors!) and for deriving the full profit from the old Astrographic Catalogue positions. Any plate of the Astrographic Catalogue could be re-reduced completely with the aid of improved positions as derived from the new proper motions and present positions. Proper motions derived from the comparison of the old plates with modern ones would have an uncertainty in the reduction to absolute of about $\pm 0''.002$ only.

For the observation of the reference stars for the repetition of the AGK 2, collaboration with other meridian observatories than Hamburg has already been sought for by Heckmann.

The conference agreed that the project proposed by Heckmann would be one of fundamental importance for future work on the proper motions of faint stars. The idea of repeating the AGK 2 in 1960, instead of in 1970–80, found strong support; it was generally admitted that the accuracy of the proper motions to be obtained would be quite satisfactory for statistical investigations, and that the special circumstances mentioned by Heckmann (the availability of experienced personnel at Hamburg) fully justify the advanced repetition.

(4) RADIAL VELOCITIES

The necessity of extensive work on radial velocities has become apparent during the discussion of many topics at the conference; it contrasts strongly with the decreasing interest shown in this type of observational work by some observatories which formerly were active in this field, and it is hoped that the present report will stimulate new activity.

The following summarizes principal programmes, to most of which we have referred already in other sections of this report:

(1) The OB super-giants found in the objective prism surveys, which will give information on the systematic motions at large distances in the galactic plane (Section A (2)).

(2) Faint Cepheids, mainly those of population I in the galactic plane which will serve for the same purpose as the OB super-giants (see Section A (2)).

(3) RR Lyrae and long-period variables to be found in the surveys described in Section A (1). Such radial velocity programmes will enable us to derive full profit from the surveys of these distant objects. They will provide information on the galactic field of force and on the kinematical differences between discrete period groups among the variables.

(4) Observations of stars in the Kapteyn Selected Areas below 10° latitude, brighter than 12·5 photographic magnitude, in a region 3·5 × 3·5 degrees (see the discussion and the list on pp. 35 and 36); in addition to these the high latitude Selected Areas mentioned in Section B (2), and also the test regions at low latitude (see p. 29).

(5) The stars in the O associations.

(6) The faint main sequence B2–B5 stars to be used for a fundamental determination of the constant *A* of Galaxy rotation (Section D (2)).

Dr Fehrenbach gave a report on the development of his method of objective prism radial velocity observations. Preliminary work with a 15 cm. refractor has furnished the necessary experience for future work, which will be carried out with a 40 cm. refractor, especially built for this purpose. Its focal length will be 400 cm. and the dispersion of the spectra, 110 A./mm. at H_γ. The magnitude limit will be 12·5 for exposures of 4 hours (2 hours for each of the two spectra to be compared with opposite directions of dispersion).

The precision of the radial velocities obtained from one plate will be of the order of ±6 km./sec. for the A stars, and somewhat higher, about ±4 km./sec., for the later types. These figures are based on the experience gained with the 15 cm. refractor.

The area measured with the small instrument was 3 × 4·5 degrees. A number of plates in low latitude areas has already been taken with this instrument. The measurements will be carried out by the Astrographic Catalogue division of the Paris Observatory.

Fehrenbach's method will prove to be extremely useful for all those programmes where a large number of stars has to be observed in a small area of the sky, like the Selected Areas. These new developments were, therefore, highly appreciated by the conference and it expressed the hope that soon much progress will be made with the measurement of the plates. It therefore adopted the following resolution:

La mesure des vitesses radiales au prisme objectif, telle qu'elle a été mise au point par M. Fehrenbach, est de la plus grande importance pour l'étude du système galactique. La conférence demande instamment que ce travail soit poussé le plus activement possible, à la fois dans l'hémisphère Nord et dans l'hémisphère Sud.

The method of objective prism radial velocities requires knowledge of the radial velocities of a few standard stars in the field observed. This presents a difficult problem, and requires the co-operation of radial velocity observers equipped with large instruments.

Various participants at the conference (Bok, Lindblad, Melnikov—for Panaeyotov—, Schalén) reported investigation of the method of objective prism radial velocities. Extension of this work with still larger instruments, which perhaps might reach stars of 14th magnitude, would be of the utmost importance, especially for the study of the kinematical properties of the distant parts of our Galaxy.

D. SPECIAL PROBLEMS

(1) STUDY OF ASSOCIATIONS

A special session of the conference was devoted to discussion of work on associations. This subject of growing interest has many aspects; some of these have been dealt with already in the preceding sections.

Nomenclature

The terms *association* and *aggregate* have been used in the past. The term *association* was accepted for use in future work. In the case of the O associations, there was some difference of opinion whether or not those associations which contain B stars, but no O stars, should also be called O associations. This latter term has been used up to now by the Soviet astronomers. There are some striking cases, for instance, that of the association around *h* and χ Persei, where the earliest type stars are B 1.

Lists of associations have been compiled by Morgan and collaborators[1] and by Markarian.[2] The listed objects are for the greater part identical, but the labelling is different. It was decided that this matter will be arranged between V. A. Ambartsumian and W. W. Morgan, who might also settle the question of the terms O and B associations.

A question which may have to be considered in the future is, which name to adopt in the cases of the small associations which resemble some of the larger open clusters in size and appearance. An example is NGC 2644, usually listed as an open cluster, but now rather considered as an O association. Morgan's suggestion of distinguishing on the basis of the diameter (smaller or larger than 10 parsecs, for instance) might prove most useful.

Internal motions

These have to be studied principally on the basis of the proper motions. Radial velocities are, of course, the only way to study internal motions of the distant associations where proper motions are unmeasurable. However, as the interpretation of the observed Doppler shift is complicated by secular changes and atmospheric phenomena, these observations have to be treated with caution. Radial velocities will also be useful for picking out the stars with large motions relative to the mass centre of the association.

Proper motions can reveal internal motions only for the nearer associations (within 1000 parsecs). The brighter O and B stars in these associations are included in the list recommended for meridian observers and mentioned in Section C (2). For the fainter stars one will have to rely on photographic proper motions, possibly in combination with the reduction to absolute motions by means of the repetition of the AGK 2 (see Section C (3)) and with spectral classifications and photometry.

For the O associations these observations may yield information on the membership of later type stars and on the amount of the internal motions.

For the T associations proper motions can be obtained at present only for the nearest objects. Many of the T Tauri stars even at a few hundred parsecs distance are already fainter than photographic magnitude 12, the limit of the Astrographic Catalogue. For these a programme of first epoch plates will have to be set up. The question was discussed, whether the 48-inch Palomar Schmidt or the Lick Astrograph Survey would be sufficient for this purpose. This has to be investigated. Cameras with a large field will be required for the study of the T associations, as these usually occur in obscuring regions where reference stars are scarce. The most suitable, nearest regions for these observations may be the dark clouds in Ophiuchus and Scorpio, the nearer clouds in the great Rift, the regions of the O associations in Cepheus, Perseus, Orion, and Monoceros, and the Taurus clouds: and in the southern sky the coal-sack and, perhaps, the Carina region.

Attention was drawn by Ambartsumian to the importance of inclusion of open clusters with O and early B stars in the programmes for proper motion. According to Markarian, noticeable internal (expanding) motions exist, for instance, in the southern open clusters IC 2602. A working list of Trapezium systems is being drawn up at the Burakan Observatory. This will be a suitable subject also for double star observers.

Physical properties of stars in associations

As was pointed out by Ambartsumian, the possibility of systematic differences between the physical properties of stars in and outside associations has to be kept in mind. Such differences occur, for instance, in the case of the O stars connected with the Orion nebula, which differ in some spectral features from other O stars. This subject should further be pursued.

According to Parenago, there are also differences in the character and number of variable stars associated with different O associations (Orion, Monoceros) which might represent different stages of evolution. Differences in the character of T Tauri stars depending on their location in HI and HII regions were noticed by Haro. According to Bok and Miss Hoffleit, there seem to be also systematic differences in the sizes of the globules in O and T associations.

Another important field for study are the light variations of T Tauri stars discovered spectroscopically in the T associations.

Absorbing matter in associations

There is evidence of deviations from the general law of reddening in the regions of high interstellar density in some associations. The situation is not quite clear as former results by Stebbins and Whitford, Schalén, and Sharpless, were not confirmed by Miss Dinant and Chalonge. This will be an interesting subject for photometric work; the nearest associations will be the most suitable as the absorption in distant associations cannot be separated from that in the nearer clouds.

(2) FUNDAMENTAL DETERMINATION OF THE CONSTANT *A* OF GALACTIC ROTATION FROM RADIAL VELOCITIES

There is still considerable uncertainty in the present determinations of this constant. These determinations were based mainly on the radial velocities of the apparently brighter O and B stars using adopted values for the absolute magnitudes; the latter were, however, rather uncertain especially in the case of the most luminous stars. A fundamental determination of the constant A based only on the stars with the most reliable absolute magnitudes is, therefore, much wanted. As a next step, we should like to try whether possible variations of the constant A as a function of galactic longitude can be detected.

A proposal for an improved determination of the constant A has been published elsewhere.[1] It is based on the fact that reliable absolute magnitudes of the main sequence B2 to B5 stars can be found from the proper motions of the brighter stars of these types ($m < 7{\cdot}0$). The constant A can then be found from the radial velocities of the faint, distant stars of apparent magnitude about 10. This programme requires the identification of the stars (which probably can be done on the plates taken already for the survey of super-giants), observations of magnitudes, colours, luminosities and spectral classifications of the individual stars, and radial velocities with probable errors below 4 km./sec.

The programme would be a rather extensive one, but the importance of a fundamental determination of the constant A in connexion with distance determinations in general seems large enough to justify the effort.

(3) TRIGONOMETRIC PARALLAXES

Trigonometric parallaxes of high accuracy are needed especially for two purposes:

Calibration of spectroscopic absolute magnitudes

With the development of new and very sensitive criteria for absolute magnitudes by means of the multi-colour photometry (see Section B (1)), the need for very accurate absolute magnitudes has also increased. They are wanted for the average relation between the measured photometric quantity and the absolute magnitude as well as for the determination of the cosmic scatter of the absolute magnitudes. For the present the first requirement would concern the F stars, for which the photometric criteria have been most accurately developed.

Absolute magnitudes and motions of sub-dwarfs

For this group, there still exists considerable uncertainty with regard to their luminosities. This is mainly due to the large peculiar motions, which make the method of parallactic motions unsuitable; particularly because of the pronounced selection of stars according to large proper motion. A particularly interesting aspect of the study of these stars is, that among them are objects of very high space velocity which in their orbit approach the galactic centre very closely,[1] and also the objects which enable us to estimate the velocity of escape from the Galaxy.

For both categories, lists of stars for which trigonometric parallaxes are most wanted will be drawn up and made available through the intermediary of the sub-commission for co-ordination of galactic research.

(4) STARS WITH LARGE PROPER MOTIONS

This subject has already been mentioned on pp. 37 and 38. Full profit can be derived from the proper motion data only when apparent magnitudes and spectral types (or multi-colour photometry) also become available. Work in this field has been done already by Luyten and collaborators,[2] but much more remains to be done. We refer to an outline of the problems involved, given by Luyten in *A.J.* **51**, 2, 1944.

In addition to these observations—which will constitute a very extensive programme—one might undertake a survey of more large proper motions, especially for the northern hemisphere. Ross' survey is complete only for proper motions exceeding 0″.4. Repetition of the Astrographic Catalogue plates might serve for this purpose, but they contain only stars brighter than 12·5, whereas for the study of the faint end of the luminosity law we are especially interested in the stars of still fainter apparent magnitudes. Perhaps the Astrographic Chart plates ($m < 14{\cdot}5$) can be used, if the multiple images on these plates do not hamper the discovery. However, full profit can be derived from the new discoveries only when a substantial number of them are observed for trigonometric parallax.

(5) INTERSTELLAR ABSORPTION: DEPENDENCE ON WAVE-LENGTH

In all statistical work where colour excesses are used as a measure for the total amount of absorption, knowledge of the dependence on wave-length is presupposed. Although there is evidence that a uniform law is applicable in the general region around the Sun, investigations of special regions or distant objects should be encouraged in order to find out how large the deviations from the uniformly adopted law can be. Work on this line has already been done by Stebbins and Whitford,[1] Schalén,[2] Van Rhijn[3] and others.

(6) INTERSTELLAR POLARIZATION

This subject was only briefly touched upon at the conference. It seemed desirable to review possible fields of research for new observers.

Measures of the interstellar polarization have been done mainly by Hiltner and by J. S. Hall. Hiltner has concentrated on observations of OB stars up to the largest possible distance in the galactic plane. His observations are supplemented by photometry and spectroscopic observations which will furnish the data on interstellar reddening and on distances, which are necessary for the interpretation of the polarization data. Hiltner has also measured some stars in clusters and associations, and is now preparing for measurements of the dependence of polarization on wave-length. Hall has concentrated, on the detailed phenomena in associations and star clusters. It is hoped that these observations will provide the basis for the study of the relation between the run of the magnetic lines of force and the spiral structure in the Galaxy and of the local deviations in the alignment of the polarizing particles.

An important field of research which is still untouched is that of the polarization in the region around the Sun up to, say, 300 parsecs. A complete scanning of this volume of space seems particularly interesting when it is supplemented with accurate observations of the distances and colour-excesses of the measured stars.

Further objects for measurement are the faint Cepheids (and other variable stars) which will supplement the observations of the OB stars, and double stars of different separations.

REFERENCES

PAGE 4

1. See, for instance:
   Van Gent, H., *B.A.N.* No. 243, 1933.
   Shapley, H., *Proc. Nat. Acad. Sci.* **25**, 113, 1939; Harvard Repr., No. 158.
   Oort, J. H. and van Tulder, J. J. M., *B.A.N.*, No. 353, 327, 1942.
   Kukarkin, B. V., 'Investigations of Structure and Evolution of Stellar Systems on the Basis of Studies of Variable Stars' (monograph); *A.J. U.S.S.R.* **24**, 269, 1947 (*Astr. Newsl.* No. 37, 1).
   Parenago, P. P., *A.J. U.S.S.R.* **25**, 123, 1948 (*Astr. Newsl.* No. 25, 6); *Progress of Astron. Sciences*, **4**, 69, 1948.
   Perek, L., *Contr. Astr. Inst. Brno*, **1**, No. 8, 1951.

PAGE 5

1. References to current variable star surveys are given below (in reference 1 to page 18).

PAGE 8

1. See also W. J. Luyten, *A.J.* **58**, 75, 1953, for an extension of Humason and Zwicky's survey.
2. *Ap. J.* **113**, 432, 1951.

PAGE 9

1. Baade, W., *Publ. Michigan Obs.* **10**, 16, 1950.
2. See the references for p. 1.

PAGE 11

1. Minkowski, R., *Publ. Michigan Obs.* **10**, 25, 1950.
2. Nassau, J. J., 'Report on the Warner and Swasey Observatory, 1951–1952', *A.J.* **58**, 269, 1953.

PAGE 14

1. Holmberg, E., *Meddel. Lund*, Ser. II, No. 128, Table 26, 1950.

PAGE 15

1. Stebbins, J. and Whitford, A. E., *Ap. J.* **108**, 413, 1948; Mount Wilson Cont. No. 753.
2. *B.A.N.* No. 452, 1954.

PAGE 16

1. *Publ. Michigan Obs.* **10**, 45, 1951; *Ap.J.* **113**, 141 and 309, 1951; *Ap.J.* **114**, 482, 1951; *Ap.J.* **115**, 475, 1952; *Bull. Tonantzintla and Tacubaya Obs.* Nos. 7 and 8, 1953, No. 9, 1954.
2. Johnson, H. L. and Morgan, W. W., *Ap.J.* **117**, 313, 1953; Cont. McDonald Obs. No. 216. See also *Ap.J.* **118**, 92, 1953.

PAGE 17

1. *Ap.J.* **78**, 87, 1933; **98**, 153, 1943; **110**, 387, 1949; **113**, 624, 1951.
2. González, G. and G., *Bull. Tonantzintla and Tacubaya Obs.* No. 9, 1954.

PAGE 18

1. The following is a list of references to descriptions of various variable star programmes.
   Harvard: Harvard Reprint No. 290, 1946 (*Ric. Astr. Vaticana* **1**, No. 11). *Harvard Annals*, **115**, 1 (Milton Fields).

Leiden: *B.A.N.* No. 397, 1948.

Sonneberg and Babelsberg: *Kleine Veröff., Berlin-Babelsberg,* Nos. 19, 24, 28; *Veröff. Sonneberg,* Nos. 2, 3, 5; *Sitz. Ber. Berlin,* 1928, 258; *A.N.* **250**, 397, 1953.

Soviet Programme: *Veränderliche Sterne* **5**, 101, 1937. Kapteyn Selected Areas, see p. 7 of this report. See also, *Trans. I.A.U.* **8**, Report Commission 27.

PAGE 20

1. *Ap.J.* **113**, 367, 1951.
2. See, for instance, Wilson, R. E. and Merrill, P. W., *Ap.J.* **95**, 248, 1942; Oort, J. H. and van Tulder J. J. M., *B.A.N.* No. 353, 1942.

PAGE 21

1. Report of the Lick Observatory, *A.J.* **58**, 251, 1953.

PAGE 22

1. *Zs.f. Astroph.* **34**, 1, 1954 (with J. Stock).

PAGE 24

1. See E. v. P. Smith and H. J. Smith, *A.J.* **59**, 193, 1954.

PAGE 25

1. Summaries have been given, for example, in B. J. Bok's *The Distribution of Stars in Space,* and *Popular Astronomy,* **52**, 261, and 318, 1944, and in the Reports of Commission 33 in the *International Astronomical Union.*

PAGE 26

1. *Publ. Michigan Obs.* **10**, 33, 1950.

PAGE 27

1. *Ap.J.* **112**, 554, 1950 and **116**, 122, 1952.
2. See also A. N. Vyssotsky and A. Skumanich, *A.J.* **58**, 96, 1953, for the extension of this work to faint stars.

PAGE 28

1. For a more detailed description see B. Strömgren, *A.J.* **56**, 142, 1951; **57**, 200, 1952; **59**, 193, 1954; *Vistas in Astronomy* (in the Press).

PAGE 29

1. The following are references to the Cleveland papers dealing with the low latitude fields; *Ap.J.* **106**, 1, 1947 (with list of co-ordinates of the low latitude fields); **109**, 139, and 414, 1949; **109**, 426, 1949 (observational data); **110**, 40, 1949; **112**, 90, 1950 (observational data); **113**, 672, 1951; **115**, 479, 1952.

PAGE 31

1. *A.J. U.S.S.R.* **30**, No. 6, 1953; *Trudi Sternberg Astr. Inst.* **26**, 1954.

PAGE 33

1. *Variable Stars,* **9**, 233 and 349, 1953.

PAGE 35

1. A list of average proper motions, derived from the Radcliffe and Pulkovo measures is kept at the Leiden Observatory. See also L. Binnendijk, *B.A.N.* No. 362, 1943, for further details.

PAGE 37

1. Publ. Obs. Minnesota, **3**, Nos. 1, 2 and 4, 1941–44; *A.J.* **55**, 15, 1949.

PAGE 38

1. Publ. Obs. Minnesota, **3**, No. 4, 1944.
2. *Ibid.*

3. *A.J.* **36**, 96, 124, 172; **37**, 53, 193; **38**, 117; **39**, 140; **40**, 38; **46**, 157; **48**, 10; 1926–1940.

4. *Veröff. Heidelberg*, Band 7, No. 10, 1919 and subsequent notes in the *Astronomische Nachrichten* Nos. 4996–5658.

5. *Ap.J.* **116**, 283, 1952; *Proc. Nat. Acad. Sci.* **37**, 637; **38**, 494; **39**, 135; **40**, 137, 1951–1954; *Proc. Am. Acad. Arts and Sci.* **81**, 255, 1952.

6. Similar lists have been published by Parenago in *Annals of the 10th All-Soviet Astr. Conference*, pp. 263–7 (Pulkovo Obs., 1954).

PAGE 39

1. For a full account of Heckmann's proposal see the Report on the Dearborn Symposium on Astrometry, *A.J.* **59**, No. 2, 1954.

PAGE 43

1. *Ap.J.* **118**, 318, 1953.
2. *Dok. Akad. Nauk Armenia*, **15**, No. 1, 1952.

PAGE 45

1. *Trans. I.A.U.* **8**, Report of Commission 33; 1954.

PAGE 46

1. See N. G. Roman, *A.J.* **59**, 307, 1954.
2. *Ap.J.* **96**, 55, 1942; **102**, 196, 1945; *A.J.* **59**, 26, 1954.

PAGE 47

1. *Ap.J.* **107**, 102, 1948; **102**, 318, 1945; **98**, 20, 1943.
2. *Uppsala Astr. Ann.* **3**, No. 5, 1951.
3. *B.A.N.* No. 441, 1953.

APPENDIX

Recommendations for Meridian Observations

List A. O and B-type Stars and Cepheids, North of Declination −20°

(See the explanation on pp. 38 and 39)

	BD		m		BD		m		BD		m
	42°	4831	8·6		68°	74	7·9		56°	498	8·7
P	62	2363	7·4		61	223	8·6	P	56	522	6·7
P	58	11	6·7	P	60	188	7·3	P	56	530	6·7
	36	12	6·6		56	240	7·6		51	548	7·0
	14	14	2·9		54	258	7·4		55	587	8·8
	50	46	6·1		59	251	8·3	P	55	588	6·8
	60	25	7·9	P	59	260	7·3		46	557	6·1
	61	38	8·0	P	62	259	7·5	P	56	568	6·5
	61	50	8·2	P	59	271	7·3		56	567	8·5
	51	62	5·4		63	218	8·0	P	55	598	5·2
	71	16	6·9		49	444	4·2	P	56	591	7·5
P	50	72	7·5–8·7		57	399	8·2	P	56	593	7·0
	57	85	7·2		54	396	5·5		40	501	7·7
	64	52	8·2		54	372	8·6	P	55	612	6·2
	61	101	7·3		54	398	7·6		58	467	8·0
	66	35	8·2		62	320	3·4	P	57	568	7·3
P	62	102	4·2		58	331	7·0		60	493	8·4
	53	105	3·7		64	268	8·0		51	579	8·0
	72	35	7·1	P	63	274	5·6		60	488	8·4
	32	101	4·4		58	356	8·2		60	487	8·2
	14	76	5·9		63	281	8·0	P	57	576	7·3
	38	91	8·2		48	600	8·0		56	635	8·5
	49	164	4·8		64	295	7·5		62	411	8·0
P	63	81	7·4	P	57	494	5·9		70	182	8·0
	51	133	6·9	P	56	438	6·4	P	57	582	7·2
	47	181	5·5		55	534	8·2	P	60	502	7·8
	47	183	4·7		58	396	8·2	P	60	504	8·0
	47	201	8·2	P	57	519	6·5		56	648	8·0
	63	97	8·4		63	310	8·0		54	569	8·0
	40	171	4·4		54	494	8·1		56	656	7·7
	58	119	8·0		55	552	8·5		59	513	8·4
P	62	160	7·1		57	525	8·5		62	424	8·5
	55	191	7·7	P	57	526	7·8		57	602	8·5
P	59	144	2·2		55	554	8·0		58	498	8·4
P	62	175	7·7		56	469	8·6		60	541	7·8
	55	216	7·8		56	470	7·2		−0	406	4·0
	61	200	8·6	P	56	471	6·4		56	693	8·4
	46	245	6·7		56	475	7·7	P	59	535	7·7
	50	212	6·5		55	564	8·6		27	424	4·6
	56	191	7·1	P	63	315	7·1	P	57	632	7·5

Recommendations for Meridian Observations (cont.)

	BD		m		BD		m		BD		m
P	57°	634	8·0	P	31°	666	2·9		43°	1147	7·3
P	68	200	6·0–6·4	P	52	726	6·7	P	−14	1003	5·9
P	59	552	7·1	P	48	1019	7·0		40	1142	3·9
	60	586	8·3	P	30	591	6·2		23	804	8·6
P	63	367	7·8		44	816	7·8		−7	948	4·8
	59	578	7·9	P	34	768	5·5		58	804	5·3
	60	608	7·0	P	39	895	3·0		44	1088	8·0
	62	504	8·0	P	35	775	4·1		21	754	8·2
	8	455	4·7		62	643	7·3		−14	1027	6·4
	57	681	8·3		53	723	7·3		41	1058	3·3
P	61	525	6·5		58	685	8·0		38	1020	7·7
	51	681	6·2		12	539	4·0		24	755	5·5
	17	493	6·1		−1	572	5·3		21	766	6·0
P	59	609	7·1		52	752	8·2		36	1021	7·8
	48	870	7·9	P	61	669	6·8		40	1189	8·4
	59	625	7·5	P	32	714	6·7		37	1067	6·2
	65	340	4·8		5	584	5·3		40	1196	8·1
	−17	631	7·8		−20	769	6·4		−8	1040	4·3
	49	899	5·3	P	61	676	7·0		41	1106	8·0
	49	902	5·1		61	676	7·1		40	1213	7·3
	20	543	5·2		51	861	7·5	P	34	980	5·8
	48	899	5·3	P	31	703	6·9	P	−8	1063	0·3
	59	648	7·7	P	33	785	6·6		57	874	6·2
	48	913	5·9		47	939	4·0		18	812	7·5
	61	587	8·7		−16	796	5·5		−17	1069	6·5
	48	920	4·9		83	104	5·4		−7	1028	3·7
P	59	660	4·4		8	657	4·3		36	1086	7·4
P	58	607	4·8		50	973	5·5		36	1090	8·2
	49	945	4·7		45	931	7·2	P	37	1146	6·7
	44	714	7·4		45	933	8·5		−1	859	6·4
P	29	566	7·1	P	53	779	5·9		41	1162	5·1
	44	734	6·3		53	779	6·6	P	−13	1127	4·3
	63	426	7·7	P	−13	893	5·5	P	37	1160	7·4
P	34	674	5·8	P	−3	834	4·1		3	857	6·4
	47	857	4·3	P	18	661	7·2–7·7	P	8	933	5·7
P	56	824	6·8		31	803	7·1		−0	929	5·7
	54	698	8·4		7	676	6·9		−0	930	4·7
	47	876	3·1		7	678	8·0	P	3	871	5·0
P	33	698	5·0		−8	929	5·9		−0	936	5·6
P	33	704	7·9		0	834	7·3		−14	1119	5·2
P	31	642	3·9		−3	876	4·2		35	1095	8·4
P	31	643	8·2	P	36	937	8·0	P	20	948	6·8
	5	539	5·4	P	66	358	4·4		2	947	6·3
P	31	649	6·5		28	704	7·7	P	−2	1235	3·4
P	33	717	6·4		8	775	7·8	P	1	1005	4·7
P	52	714	6·8	P	5	745	3·8	P	6	919	1·7
	10	486	5·0		9	668	6·1		16	775	6·2
P	52	715	6·9	P	2	810	3·9		−1	889	7·3
P	33	728	5·7	P	35	930	6·2		0	1056	6·0
P	33	730	7·5		0	893	5·9	P	30	898	5·7

	BD	m		BD	m		BD	m
	34° 1046	9·0		−4° 1184	6·3		27° 914	7·7
	34 1049	8·6	P	−4 1183	6·5		−4 1281	6·4
	34 1051	8·4	P	−4 1185	4·7	P	24 1033	6·0
P	33 1049	7·5	P	−5 1315	5·4		16 926	6·9
	17 928	5·3		34 1118	8·1	P	25 1052	4·9
	21 847	4·8		−4 1186	6·3		28 991	7·9
	2 962	4·7	P	−5 1319	5·2		46 1091	7·0
	3 901	7·7		−5 1320	6·5	P	20 1233	4·7
	3 903	6·6	P	−6 1241	2·9		48 1339	6·5
	−2 1250	6·6		−3 1146	6·3		−6 1391	5·1
	−1 897	7·7		−4 1187	7·4		28 1008	7·5
	−2 1254	7·4	P	−1 969	1·8		62 818	8·6
	1 1021	6·4		33 1103	8·3	P	21 1120	8·0
P	35 1137	6·7		−4 1190	7·0		−4 1362	5·4
	1 1026	7·6		−5 1334	6·5		14 1152	4·4
	3 928	7·5		21 908	3·0		−11 1386	6·4
	−7 1092	6·6		−6 1255	5·6		13 1124	7·4
	1 1032	5·7		8 1016	6·1		15 1079	8·5
	34 1077	8·0		−6 1262	5·8	P	23 1226	5·8
	11 834	7·3		28 836	8·2		20 1284	7·4
	5 939	4·3		−5 1342	7·3		−5 1521	8·4
	−7 1099	6·2		−4 1196	6·3		13 1147	6·7
	3 944	9·0		29 947	6·0		21 1143	7·8
	3 948	5·5		−1 987	6·7	P	20 1302	6·9
P	32 1024	4·9		25 902	5·0		16 1035	4·9
	1 1045	7·9	P	−2 1326	3·8		14 1187	4·4
	−6 1207	6·0		−2 1327	6·5		7 1178	6·9
	26 841	8·2		−6 1275	5·9		10 1044	7·4
	36 1177	7·7	P	4 1002	4·5		−6 1446	5·1
	−0 982	6·9		−5 1351	7·4		−14 1359	8·1
P	−0 983	2·5		38 1250	8·0		8 1238	8·5
P	−7 1106	4·6		−11 1251	8·0		−17 1398	6·3
	18 877	5·5		42 1376	7·0	P	13 1173	5·8
	35 1169	8·4		16 841	4·9		4 1181	6·4
P	−1 935	5·3	P	−2 1338	2·1	P	23 1275	6·3
	−4 1164	8·0		−2 1337	6·1		23 1297	8·1
	−1 939	6·5		−1 1004	5·0	P	23 1300	7·0
	14 947	5·6		−1 1005	8·2		20 1369	8·4
P	−1 943	5·4		−2 1345	8·6		−19 1407	5·3
	1 1058	6·4		23 1015	6·1		−7 1373	5·1
	−1 949	6·2	P	25 941	6·9		11 1128	6·4
	23 954	5·3		30 992	8·1	P	−11 1460	5·5
P	9 877	4·5		13 979	5·2		15 1176	7·7
	5 958	6·7		36 1261	8·2		25 1251	8·5
P	9 879	3·7		28 902	8·5		3 1221	6·3
	21 899	7·8	P	−9 1235	2·2	P	−17 1467	2·0
	9 881	7·7		−7 1187	5·3		11 1162	7·7
	−0 1009	8·1		26 985	8·1	P	7 1273	5·8–6·8
	−6 1233	5·6		25 1019	8·5		14 1296	7·1
P	−6 1234	4·7		19 1126	5·9		−4 1510	6·1

	BD	m		BD	m		BD	m
P	30° 1238	5·0–5·6		0° 1691	8·4		−13° 2143	8·3
	−4 1526	5·0		24 1457	7·7		43 1754	7·0
	−12 1500	7·5		−1 1446	6·3		3 1848	7·7
	−6 1574	4·7		−16 1661	4·4		−1 1900	8·1
	7 1314	7·9		18 1423	7·1		−2 2379	6·4
	2 1262	8·8		−4 1745	8·5		−4 2197	7·4
	5 1267	6·7		−8 1639	8·3		6 1867	8·1
	4 1282	7·9		−10 1774	7·0		−19 2228	6·1
	11 1204	5·8		1 1610	7·9		−18 2190	4·3
	4 1291	8·0		−2 1856	7·7		−15 2280	5·5
	−13 1519	6·1		−12 1729	7·9		−13 2429	8·5
	5 1279	8·1		−5 1912	7·0		−14 2526	6·6
	5 1282	7·7		−8 1667	6·4		3 2039	4·3
P	5 1283	6·8		−2 1885	7·9		−13 2718	8·2
	5 1286	8·2		5 1514	6·5		50 1608	7·5
P	4 1302	7·1		−11 1747	6·6		79 300	8·1
P	7 1337	4·5	P	−4 1788	4·9		−13 2917	5·0
	−7 1462	8·4	P	20 1687	3·7–4·1		−2 2986	8·5
	4 1319	7·7	P	−10 1848	7·0		22 2164	5·6
	4 1318	8·3		−11 1770	7·8	P	17 2171	3·6
	2 1295	8·1		−8 1734	7·8		3 2352	6·5
P	15 1246	6·7–7·5		−10 1862	6·4	P	10 2166	3·9
	11 1232	7·9		−12 1777	7·2		38 2179	6·9
P	10 1193	7·8	P	−11 1790	5·3		−2 3312	7·3
	6 1303	7·3		−12 1788	6·4		17 2374	5·8
P	6 1309	6·1		−12 1809	7·0		−18 3295	5·3
P	5 1334	6·2	P	−10 1892	6·2	P	−10 3672	1·2
	−8 1498	8·5		−16 1802	6·0	P	−17 3918	7·3
	4 1360	8·3		−11 1822	7·9		−1 2858	7·7
	4 1361	7·9		−15 1695	8·1		50 2027	1·9
	4 1363	8·4	P	−10 1933	6·0		−14 4160	8·3
	7 1386	7·4		−10 1934	8·4		15 2862	7·9
P	1 1443	6·1		−19 1767	7·3		−16 4110	5·6
	−4 1607	8·5		−15 1732	6·8		−19 4249	5·1
	9 1334	7·9		−11 1867	8·5		−18 4195	5·9
P	10 1220	4·7		−5 2080	6·6		−13 4302	4·7
	9 1344	7·0	P	−8 1872	6·2	P	−19 4307	2·9
	−12 1585	6·8		15 1564	6·4		−19 4308	5·1
P	6 1351	6·2		−19 1854	7·3		−19 4333	4·3
	−5 1753	6·9		−15 1810	5·2		46 2169	3·9
	1 1472	7·8		2 1677	8·3		−18 4282	4·9
P	4 1414	5·8		−9 2069	6·6		−5 4318	7·9
	0 1574	7·7		−13 2051	6·9	P	−10 4350	2·7
	67 454	5·0		−5 2148	8·7		14 3086	6·6
	2 1379	7·5		−15 1892	8·3	P	−0 3224	5·6
	8 1486	5·8	P	−14 1966	6·2		45 2509	7·4
P	1 1531	6·1		−19 1950	6·8		1 3408	6·0
	−5 1815	7·3		17 1623	7·7		33 2864	4·8
	−13 1682	7·9		−14 1999	5·6		−10 4493	7·4
	5 1448	6·8		−19 1967	5·7		−17 4834	8·2

Recommendations for Meridian Observations (cont.)

	BD	m		BD	m		BD	m
	46° 2349	3·8		−17° 5149	8·3		−19° 5242	7·0
	−7 4487	6·2		−17 5151	8·0		19 3848	8·0
	−6 4638	Var.		−18 4926	5·4–6·2		17 3799	5·8–6·2
	−9 4598	8·3		5 3704	6·0		−13 5172	5·4
	−2 4458	7·9		24 3395	6·9		20 4007	6·7
	34 3050	6·6		−16 4850	8·3		9 3951	6·5
	15 3285	8·0		−9 4713	8·5		40 3544	6·1
	−6 4672	6·2–7·0		−14 5037	8·2		21 3634	6·9
	0 3813	5·7	P	−14 5039	6·8		20 4022	6·6
	−14 4842	8·0		−9 4729	8·5		−10 4926	6·7
	4 3570	4·8	P	−9 4736	7·8		26 3429	5·5
P	2 3458	3·9		22 3358	6·7		50 2708	5·2
	6 3597	6·2		26 3255	6·8		33 3295	6·2
P	−19 4800	7·3		26 3257	6·9		9 3979	6·9
	20 3649	5·1		−16 4888	8·4		23 3549	6·9
	−15 4803	8·1		26 3259	6·4		3 3902	7·8
P	1 3578	6·1		−13 5003	8·5		−19 5312	5·4
	−14 4880	8·2		3 3727	6·5		41 3232	6·2
	−16 4720	7·4		−16 4907	8·4		1 3899	7·0–7·7
	20 3674	4·3		23 3347	5·7		34 3437	6·6
	−16 4736	7·6		−5 4678	7·3		−8 4887	5·4
	−16 4737	7·6		−10 4713	5·8		9 4037	8·2
	−16 4744	8·2		−19 5047	6·5–7·3		38 3490	4·5
	−16 4747	8·0	P	−15 5004	8·1		21 3713	4·6
	−15 4856	7·9	P	−18 4994	7·0		0 4159	8·1
P	−19 4895	7·1		30 3227	6·4	P	22 3648	5·4
	−16 4752	7·3		−18 5008	7·1		23 3625	7·5
	−19 4900	8·4		−14 5131	8·3		33 3409	6·3
	−15 4868	7·9		34 3245	5·9		31 3544	6·6
	−18 4844	8·4		50 2618	7·4		−8 4950	6·5
	−19 4917	8·5		−0 3523	8·1		25 3802	7·3
	−18 4857	7·9		−8 4675	7·9		25 3803	7·3
	−19 4928	7·0		−15 5063	8·3		22 3674	7·7
	−12 4953	8·3		34 3285	6·1		22 3675	7·9
	−12 4954	8·4		−1 3553	8·1		29 3584	4·9
	−17 5092	8·2		−7 4689	7·9		−2 4998	8·2
	−18 4871	7·9		−17 5310	7·1		21 3782	7·2
	−18 4873	8·2		−8 4714	7·6		−15 5362	6·7
	−16 4786	8·2		−16 5041	7·8–8·6		13 4020	6·3
	−19 4944	7·3		31 3369	5·8		19 4028	7·2
P	−18 4886	6·4		52 2280	5·8		−7 4968	6·3–7·0
	−19 4953	7·6		23 3465	6·5		37 3465	6·4
P	−15 4911	6·6		15 3583	6·5		19 4039	6·4
P	−12 4980	7·3		32 3227	5·8		26 3566	8·0
P	−18 4896	6·9		33 3224	7·8		−16 5337	7·3
	−13 4925	8·5		−11 4786	7·9		34 3590	4·9
	−13 4926	8·5		37 3262	7·1		3 4065	6·8
P	−12 4988	8·5		−15 5143	5·0		15 3866	6·8
	−14 4991	8·2		36 3307	5·5	P	−7 5006	5·0
	13 3593	6·2		0 4055	7·7		30 3645	7·4

Recommendations for Meridian Observations (cont.)

	BD	m		BD	m		BD	m
	20° 4200	6·5–7·6		39° 4033	8·2		38° 4010	8·0
	16 3928	7·4		23 3896	5·1		42 3692	7·7
	3 4097	6·4		35 3964	8·3	P	37 3892	7·6
	5 4225	5·2		35 3966	8·3		43 3571	6·8
	40 3824	7·5		35 3970	7·1		38 4032	8·0
P	20 4218	6·4		10 4189	6·2		40 4132	8·4
	13 4098	5·8		25 4116	7·8		44 3439	7·8
	17 4059	7·9		39 4049	8·4	P	40 4150	7·1
	22 3784	6·5		37 3783	8·0		38 4057	8·4
	10 4036	7·4		42 3599	7·9		37 3916	5·7
	46 2765	8·3		36 3896	8·5		40 4164	7·7
	−3 4701	6·5		35 3987	8·1	P	40 4165	7·5
	28 3460	6·4–7·1		35 3994	7·7		54 2348	7·2
	23 3767	8·2		36 3907	4·8		41 3758	7·2
	31 3765	7·3		35 3995	7·8		41 3765	7·9
	44 3236	7·3	P	−9 5382	6·5		36 4095	7·7
	19 4162	7·5		28 3645	6·9		48 3142	4·9
	−1 3834	8·3		35 4001	7·8	P	43 3630	7·2
P	33 3602	6·4	P	21 4088	6·1		31 4126	7·6
	7 4252	6·4		35 4006	7·8		47 3136	6·8
	22 3833	4·9		40 4050	8·2	P	42 3778	6·4
P	40 3902	5·6		41 3642	7·9		32 3862	7·1
	27 3536	7·7–9·5		47 3038	Var.		40 4227	8·5
	0 4337	3·7–4·4		35 4013	7·9		6 4576	6·9
P	18 4276	6·3		37 3821	7·4		20 4629	6·3
P	46 2793	5·5	P	39 4082	7·5		23 4084	5·0
P	47 2939	5·7	P	38 3956	7·1		15 4220	5·9
P	47 2945	6·2		38 3958	8·2		45 3233	6·5
	40 3931	6·8		35 4026	7·1	P	44 3541	1·3
	57 2084	5·0		29 3948	6·9		34 4127	6·5
	16 4067	5·5–6·1		67 1235	6·8		35 4229	8·2
	40 3948	7·2	P	36 3958	7·0		49 3353	5·4
	35 3878	6·0		36 3956	7·9		49 3352	8·1
	39 3968	5·4		39 4096	7·7		35 4234	5·7–7·3
	37 3703	6·3		25 4165	4·8		78 716	6·8
	19 4236	7·5		41 3675	8·5		31 4204	8·2
	36 3806	5·2		40 4086	7·9		42 3914	8·3
	41 3569	7·7		37 3860	8·0		56 2477	6·4
	21 4027	6·6		31 4018	7·0	P	45 3291	4·9
	45 3044	8·3		38 3980	8·1		53 2495	8·0
	35 3930	6·7		37 3866	8·0		46 3067	6·5
	36 3841	8·2		36 3987	8·0		27 3890	5·4–6·3
P	31 3925	5·7	P	37 3867	7·1	P	63 1663	6·4
	35 3949	7·8		25 4189	6·8		37 4076	7·0
	35 3953	7·0		37 3871	4·9	P	32 3974	6·4
P	35 3952	7·3		30 3980	8·3		34 4184	7·1
	35 3955	7·3	P	40 4103	5·8	P	54 2429	8·3
	46 2846	8·5	P	37 3879	7·4		42 3894	8·5
	35 3957	8·0	P	38 4006	7·3		43 3755	4·7
	34 3871	7·9		45 3139	6·3		27 3909	6·4

	BD	*m*		BD	*m*		BD	*m*
P	48° 3242	7·1		51° 3112	7·5		53° 2897	6·6
	55 2486	7·6	P	56 2617	5·6		64 1672	8·4
	39 4368	7·0		50 3410	4·8	P	42 4420	4·5
P	46 3111	5·8	P	57 2374	7·0		69 1257	7·2
P	44 3639	6·0		62 1973	8·3		59 2536	8·0
	56 2515	6·1	P	61 2193	6·0	P	40 4854	7·0
P	46 3133	4·9		60 2288	4·5		−17 6554	6·7
P	45 3364	5·2		61 2194	7·6	P	38 4808	6·6
	67 1283	7·2		51 3144	7·5	P	38 4808	5·8
	45 3384	8·1	P	48 3504	4·3	P	49 3903	6·2
P	54 2470	7·2	P	59 2420	7·0	P	37 4631	6·8
	30 4318	5·7–5·9		19 4793	6·2	P	38 4817	7·4
	14 4544	6·9	P	40 4648	6·5	P	36 4898	6·7
	32 4060	7·8	P	52 3043	6·6	P	38 4826	4·9
	45 3427	7·5		25 4635	5·1		23 4592	7·3
	44 3718	6·5		61 2209	8·1	P	39 4912	5·2
	38 4372	7·7		62 1992	7·7	P	37 4670	6·2
P	35 4426	6·4	P	62 1994	6·8		64 1704	6·8
	40 4432	7·3	P	61 2216	7·1		16 4814	7·2
	43 3842	7·8	P	55 2644	6·0	P	64 1717	6·8
P	59 2334	5·7		64 1607	5·9	P	47 3931	8·1
	43 3850	7·7		64 1611	7·6	P	41 4623	5·8
	45 3456	7·5	P	60 2320	6·9		58 2492	7·2
	37 4235	7·3		65 1691	6·3	P	61 2356	8·4
	47 3348	6·3		32 4310	8·1	P	49 3965	8·0
P	38 4431	4·3	P	6 4940	6·0	P	42 4529	7·8
	78 744	7·0		59 2443	8·0	P	42 4538	7·7
	34 4371	4·4	P	61 2233	6·5		61 2370	8·3
P	57 2309	6·4		60 2329	7·9	P	40 4949	5·5
P	43 3877	5·1	P	57 2441	5·5	P	62 2136	7·8
P	61 2112	6·6		52 3088	8·4		47 3985	5·2
	64 1527	5·2	P	59 2456	6·7		61 2373	7·7
	9 4793	8·3	P	47 3692	6·2	P	38 4904	6·1
	13 4692	6·7	P	61 2246	5·2	P	43 4355	7·0
	40 4503	7·4		−19 6227	5·7	P	62 2146	7·4
P	46 3294	7·1		60 2348	7·3		62 2147	7·7
P	54 2533	8·0		51 3281	7·6	P	37 4744	6·4
	60 2233	7·6	P	58 2402	5·2		41 4664	3·6
P	36 4557	5·8	P	45 3879	8·5	P	43 4378	6·3
	36 4568	5·2		62 2061	8·5		59 2629	6·9
	45 3549	6·9		5 4998	5·4		3 4818	4·6
	43 3941	7·5		51 3341	7·1		62 2161	7·8
P	58 2272	8·3		11 4784	4·9		62 2162	8·4
P	69 1173	3·3		54 2756	8·0		57 2689	8·1
P	59 2395	5·5	P	0 4872	4·6		63 1928	7·8
P	56 2589	7·4	P	36 4835	6·4		46 3931	8·0
	29 4453	8·2	P	39 4841	6·1	P	62 2170	7·5
	−20 6251	4·7		62 2081	8·5	P	50 3946	7·2
P	61 2169	4·9	P	64 1664	5·7	P	58 2545	4·9
	56 2614	7·6		57 2548	3·7–4·4	P	45 4147	6·6

Recommendations for Meridian Observations (cont.)

	BD	m		BD	m		BD	m
	58° 2546	6·3		54° 3006	7·4		55° 3051	8·5
	62 2171	6·2		67 1555	8·1	P	58 2676	8·4
P	48 3950	6·5		61 2509	8·5		45 4381	6·5
	52 3383	7·1		65 1943	5·9	P	54 3082	4·9
	4 4985	6·9		62 2296	8·0		60 2656	7·4
	58 2565	8·5		46 4169	5·8		59 2813	8·5
	63 1962	8·3		61 2526	8·5	P	62 2356	6·3
	63 1964	8·3		61 2533	5·6		54 3103	7·6
	60 2521	6·8		61 2537	8·2		61 2585	8·2
	55 2942	8·5		62 2313	8·5		53 3280	7·6
	60 2522	8·0		56 3106	8·3		57 2855	6·5
	−10 6098	7·5	P	60 2636	7·0			
	56 2999	6·8	P	60 2637	7·6			
	35 5024	6·8	P	61 2562	7·2			
	57 2748	4·9	P	56 3115	6·1			

Printed in the United States
By Bookmasters